Praise for
Unstable Ground

"Climate change, the critical defining catastrophe of our time, is simultaneously a scientific and a human concern. In this pathbreaking volume, Alvarez has combined a number of approaches to the core issue in order to show just what this will portend for the future—a future in which violence, genocide, and population collapse are entirely likely unless the means can be found to address the slide toward disaster. This is a thought-provoking and terrifying book that nonetheless offers us some measure of hope . . . if only we pay heed to its message."

—Paul R. Bartrop, Florida Gulf Coast University

"*Unstable Ground* provides a guide to understanding the many and varied implications of climate change—including environmental destruction, mass migration, and dissolution of established borders; the need to rethink issues of national security; and the existential question of life on planet Earth—if there is not enough action now to stem the self-inflicted wound. Alvarez's timely book is essential reading for citizens, policy makers, and scholars."

—Roger W. Smith, College of William & Mary

"Many of the major human rights issues of our time, including migration, armed conflict, and access to water, have important links to climate change, as Alex Alvarez shows in this compelling account. He connects many of the dots to explain the origins, at least in part, of the scourge of genocide and crimes against humanity today."

—William A. Schabas, Middlesex University, London

"In the highest tradition of the public intellectual, Alex Alvarez has produced a first-rate research work that is accessible to readers at every level and puts focus on a crucial dimension of human-caused global warming. His compelling analysis is that the dominant form of mass human violence in the twenty-first century will be (and already has been) driven by climate change. His book makes clear that the question facing us now is not whether these pressures will come, but how humanity will face them. This book is an absolute must-read for all policy makers, concerned citizens, and scholars."

—Henry C. Theriault, Worcester State University; founding co-editor of
Genocide Studies International

Studies in Genocide:
Religion, History, and Human Rights

Series Editor: Alan L. Berger,
Raddock Family Eminent Scholar Chair of Holocaust Studies,
Florida Atlantic University

Genocide is a recurring scourge and a crime against humanity, the effects of which are felt globally. Books in this series are original and sophisticated analyses describing, interpreting, and articulating lessons from historical as well as current genocides. Written from a range of scholarly perspectives, the works in this series articulate patterns of genocide and offer suggestions about early warning signs that may help prevent the crime.

Jihad and Genocide, by Richard Rubenstein
Balkan Genocides: Holocaust and Ethnic Cleansing in the Twentieth Century, by Paul Mojzes
Native America and the Question of Genocide, by Alex Alvarez
The Genocide Contagion: How We Commit and Confront Holocaust and Genocide, by Israel W. Charny
Unstable Ground: Climate Change, Conflict, and Genocide, by Alex Alvarez

UNSTABLE GROUND

Climate Change, Conflict, and Genocide

Alex Alvarez

ROWMAN & LITTLEFIELD
Lanham • Boulder • New York • London

Published by Rowman & Littlefield
A wholly owned subsidary of The Rowman & Littlefield Publishing Group, Inc.
4501 Forbes Boulevard, Suite 200, Lanham, Maryland 20706
www.rowman.com

Unit A, Whitacre Mews, 26-34 Stannary Street, London SE11 4AB, United Kingdom

Portions of the discussion of the role of borders and refugees in chapter 5 first appeared in earlier version in: Alex Alvarez, "Borderlands, Climate Change, and the Genocidal Impulse," *Genocide Studies International* 10, no. 1 (Spring 2016): 27–36. © 2016 Genocide Studies International. doi: 10.3138/GSI.10.1-3. Reprinted with permission from University of Toronto Press. (www.utpjournals.com).

British Library Cataloguing in Publication Information Available

Library of Congress Cataloging-in-Publication Data Available

ISBN 978-1-4422-6568-4 (cloth : alk. paper)
ISBN 978-1-4422-6569-1 (electronic)

♾ ™ The paper used in this publication meets the minimum requirements of American National Standard for Information Sciences Permanence of Paper for Printed Library Materials, ANSI/NISO Z39.48-1992.

Printed in the United States of America

For Ingrid, Joseph, and Astrid

You are all a constant source of pride,
love, and inspiration.

Each of you has enriched my life immeasurably.

CONTENTS

ACKNOWLEDGMENTS

As ever, this book could not have been written without a great deal of personal and professional support, for which I am profoundly grateful.

I would like to begin by acknowledging a number of friends and colleagues who have read early drafts and provided valuable feedback that helped me tremendously as I struggled to make sense of these issues and craft a coherent manuscript. Specifically, I wish to extend my deep appreciation to Janine Schipper, Mike Costelloe, Herb Hirsch, Sarah Prior, and Bjorn Krondorfer for reading through various chapters and offering their reactions, thoughts, and advice. I especially want to say thank you to Maureen Hiebert for a very thorough reading of the entire manuscript and for providing me with such insightful and helpful comments.

This is the second book I have completed with the team at Rowman & Littlefield, and as last time, I am impressed by their skills and professionalism. It's been a pleasure. Thank you to Alan Berger for his support, gratitude to Jehanne Schweitzer for her assistance in the production process and Cheryl Brubaker for her skills as a copyeditor, and a special note of appreciation to my editor, Sarah Stanton, for all of her hard work, guidance, and support on behalf of this book.

I also want to extend my undying gratitude to my best friend, confidant, and life partner, Donna Mae Engleson, for the endless support, encouragement, and patience as I wrestled with the subject matter of this book and the process of writing. My deepest love also to my children, Ingrid, Joseph, and Astrid, who remain my inspiration, my hope, and a never-ending source of pride.

INTRODUCTION

Climate Change and Genocide

There's one issue that will define the contours of this century more dramatically than any other, and that is the urgent threat of a changing climate.

—Barack Obama[1]

Every year we await the rains. I live and work in Flagstaff, Arizona, a community nestled at the foot of the San Francisco Peaks in northern Arizona. Situated at 7,000 feet on the Colorado plateau, Flagstaff is a beautiful mountain town that sits in the middle of the largest ponderosa pine forest in North America. Just north of town are the San Francisco Peaks, a volcanic mountain range sacred to many Native Americans. The highest peak is Mount Humphreys, at a height of more than 12,000 feet. When people think of Arizona they tend to imagine sandy deserts and giant saguaro cacti, but Flagstaff and the surrounding area don't match that imagery; in many ways, this area is an island of green in the high mountains. Go in any direction and pretty soon you drop off into another world, a drier and much more arid one.

The forest surrounding Flagstaff is full of canyons, meadows, cliffs, and other natural wonders. It truly is a beautiful area, but living here comes at a cost. Our location, as magnificent as it is, also means that fire is a constant threat, especially in the early summer months when the surrounding forest warms up and dries out. A hundred years of suppressing fires has meant that the fuels—undergrowth, dried needles, and dead wood—have accumulated to such an extent that when fires break out they can quickly grow into all-consuming conflagrations. Small fires are actually helpful since they serve to clear out the brush and debris while allowing big, mature trees to survive such smaller outbreaks. When modern fires break out, however, they tend to be so large and hot that they burn everything, even the big, old-growth trees. These fires can be catastrophic.

In 2014, a human-caused fire broke out in scenic Oak Creek Canyon, a verdant and lush gorge just south of town. While the south end of the canyon opens up to the Red Rocks of Sedona, the north end of the canyon points like an arrow toward Flagstaff and this particular fire quickly spread upslope toward my community and forced the evacuation of a couple of neighborhoods on the side of town closest to Oak Creek. Ultimately, fire crews stopped the fire before it reached town, but it was a near run thing. It wasn't the first time Flagstaff has been threatened with fire, and it certainly won't be the last. Such an event is symptomatic of the American West, which, in the last few years, has seen quite a few wildfires and every year the problem just seems to be getting worse. Drought in California and other western states has helped fuel wildfires across the West that have resulted in thousands of acres burned, homes and structures lost, and many deaths. Sometimes it feels as if the entire West has been burning. Research reveals that this isn't just alarmism. A recently published study found that climate change was responsible for a near doubling of forest fires in the American West.[2]

When living in Flagstaff one becomes acutely aware of the role that climatic conditions have on our lives and communities. The risk of fire is just one of a number of climate issues that sometimes dominate life out here. Water is another. As the population of the American West has grown dramatically in the last twenty years, access to water has often become a critical issue. Between 1980 and 2006, the western states of the United States grew over 60 percent, with most of that growth occurring in California, Arizona, Washington, Colorado, and Nevada.[3] Such growth has meant that these states have seen a large increase in demands on existing water supplies and the desire to acquire more, all of which has resulted in a great deal of conflict between and within states for access to and control of what is a scarce resource in the largely arid Southwest. This is nothing new, however, as anyone who knows the history of the native peoples from this region can attest to. The Sinagua, Hohokam, Chacoan, and many other ancient peoples coped, adapted, and sometimes struggled to survive in a landscape prone to recurrent periods of drought and in doing so developed many ingenious adaptations that included agriculture, extensive irrigation canals, widespread trade networks, traditions of cooperation and communal living, and dry-farming techniques among other sophisticated adaptations. What's old is what's new. Climate, as we shall see, has always been changing and has always impacted human communities and civilizations, with sometimes disastrous results. As we will discuss in this book, climate change has helped give birth to many civilizations, but has also played a role in their downfall. The question about climate change is not whether or not it is happening. It is.[4] Nor is the question about whether we caused or accelerated it. We have. To my way of thinking, the important question is how will humanity cope and adapt to the new climate realities this time around?

I am someone who has spent years studying the origins and nature of violence generally, and genocide specifically, and because of the work I do it was perhaps inevitable that my growing awareness of climate change would lead me to begin considering the role that climate change will play in helping create certain kinds of conflict—especially war and genocide. In a nutshell, that is the focus of this book. Specifically, this project is all about exploring, discussing, highlighting, speculating, and warning about how the impacts of climate change may foster the development of violent reactions and solutions to problems brought about by a warming world. Change is always stressful, both for individuals and communities, and as our world increasingly warms and as traditional weather patterns change, societies all around the world must confront this new environmental reality and the consequences it will have on our political, social, and economic structures and institutions. Some of the consequences of climate change will be sudden, catastrophic, and immediately disruptive, while others will be slower and only gradually make their impact known. Some regions will be harder hit than others and some regions and nations will have more vulnerability than others due to issues such as wealth, infrastructure, and geography. This book is concerned with tracing some of the ways in which these challenges and altered circumstances will heighten the risk for the development of communal and ethnic violence, war, and genocide. As we will discuss in the coming chapters, violence, both collective and individual, can be perpetrated as an expression of emotions such as fear, anger, and resentment. Such "expressive" violence is a manifestation of internal sentiments and feelings. Other times, violence can be a means to an end. This "instrumental" violence can be understood as a tool or method used to achieve a goal. It is my contention that the coming years will see a heightened risk for both expressive and instrumentalist forms of individual and collective violence being perpetrated for the specific reasons I outline in the following chapters.

While a book such as this is by necessity hypothetical and somewhat speculative, it is nevertheless grounded in what we do know about states, borders, conflict, war, and genocide. The scholarship of conflict, violence, and genocide has much to teach us about the ways in which states and communities often react when confronted with problems, crisis conditions, and threats. Over the years, a well-established body of research has developed that can provide a great deal of insight into the etiology and dynamics of social conflict and violence, the connection between resources and intergroup competition and conflict, and how and why genocidal ideologies and practices develop and are sometimes implemented. It is my goal in this book to apply this scholarly literature in ways that allow us to better understand how the consequences of climate change increase the risk of violent conflict, especially war and genocide.

In many ways, this book represents my own imperfect attempt to make sense of some very complicated issues. Each of the issues I discuss in this book is extraordinarily complex and, by necessity, I've had to make some hard choices in terms of what to include and exclude and how to approach and discuss what I see as the most relevant and important concerns. My approach to this topic is also a reflection of my own idiosyncratic means of thinking about these issues and making connections in ways that make sense to me, and which I hope resonate with the reader as well. In doing so, I have drawn from a wide range of both historic and contemporary events around many issues that I feel can shed some light or help us better understand the relationship between climate change and the potential for violent conflict. For some of the examples that I rely upon, the connections will be immediately evident, while in others they may be a little less clear, but hopefully only initially, since I have attempted to make the implicit connections more explicit in the course of the discussion. I find that the past is often helpful for understanding the present since human beings and the communities they create often react in fairly predictable ways. Kurt Campbell and Christine Parthemore put it this way: "Even though global warming is unprecedented, many of its effects will be experienced as local and regional phenomena, suggesting that past human behavior may well be predictive of the future."[5] This is the approach I adopt here in trying to explain how climate-change-induced stressors may potentially increase risks for various forms of violent conflict. The specific structure of the book is as follows.

In the first chapter, "Making Sense of Climate Change," I introduce the reader to the issue of climate change. In addition to reviewing the long history of climate change on this planet and some of the reasons why the earth's climate has varied in the past, I also explore the nature of human-caused climate change in the present era and its potential impact on weather, plants, and animals, and importantly for purposes of this book, on human societies. I make the point that human societies have always had to deal with climate changes that have often dramatically affected human communities around the world. What distinguishes the modern era of climate change, however, is that the impacts and challenges are expected to mount and accumulate over time. Rather than a relatively short-lived alteration in a local or regional climate system, we are looking at a global phenomenon that will endure a very long time and pose challenge upon challenge, potentially eroding the ability of nations to cope and adapt.

This is followed by chapter 2, "On the Origins of Violent Conflict: War and the Genocidal Impulse," in which I examine the ways in which climate-change-related issues in the past have contributed to conflict and violence between communities. I begin with a discussion of the Chacoan peoples of the American Southwest to illustrate this reality since the history of this Native American culture illustrates well the ways in which environment and climate can shape and foster the development of civilization but also lead to societal collapse, violence, and warfare. This chapter also defines and explores a number of different yet related forms of collective conflict including riots, pogroms, warfare, and genocide to assist the reader in understanding the mechanisms and processes that make these forms of violence possible. While there are many more types of collective violence, I focus on these forms since I believe these examples most clearly illustrate the violent potentials arising out of climate change. I also suggest that they represent various points along a spectrum or continuum of violent conflict that range from the most spontaneous and least organized (riots and pogroms) to the most planned and organized (war and genocide).

I begin chapter 3, "Linking Climate Change and Conflict," with a discussion of the Bosnian experience with genocide and ethnic cleansing in the 1990s, even though this example was not created by climate change. I nevertheless rely on this particular case since it very clearly illustrates the processes and mechanisms through which an apparently stable society can descend into war and genocide as a result of bad leadership, social, political, and economic instability, and a resurgence of ethnic nationalism, prejudice, and xenophobia. With Bosnia as the springboard, I then identify and examine the specific pathways linking climate change and violent conflict through the development of certain specific situational, ideological, and psychological conditions. This assessment includes discussion of things such as state failure, resource scarcity, scapegoating, and the persecution of marginalized populations.

This discussion is followed by chapter 4, "Water, Violent Conflict, and Genocide," in which I examine one particular resource that is expected to be dramatically impacted by climate-induced change, and that is water. For a number of reasons that are outlined in this chapter, some regions will experience flooding from predicted sea level rise, while other regions will suffer from water scarcity. This chapter explores the various ways in which flooding and drought will increase the risk of conflict and genocide, and in fact, have already helped facilitate war and genocide in places such as the Darfur region of the Sudan, where the first genocide of the twenty-first century has been perpetrated, and Syria, where an uprising that degenerated into civil war and helped spawn the rise of ISIS was partially facilitated by a drought that profoundly impacted and destabilized Syrian society.

Chapter 5, "Forced Displacement and Borders in a Warming World," is concerned with the ways in which nations and their borders will be confronted by large numbers of people dispossessed of their homes both directly and indirectly by climate change. The influx of climate change refugees will challenge the ability of nations to meet the needs of these displaced persons and may well serve as a source of friction and conflict over increasingly scarce resources as vulnerable refugee populations become sources of tension, blame, anger, and fear. Furthermore, I also discuss the issue of border regions as potentially volatile zones of conflict because of the physical and spatial geography of these areas. Finally, I examine the case of Bangladesh (which because of its low-lying topography, population density, and economic vulnerability is particularly at risk for population displacement) and explore its risks for persecution and conflict, especially with their Indian neighbors.

Finally, in chapter 6, "Preventing Conflict and Building Resilience," I discuss a number of important concerns and concepts relative to policy options, intervention strategies, and other responses to the challenges and stresses of climate change. While not definitive, this chapter is intended to provide the reader with some important issues that must be considered when thinking about reducing or defusing the kinds of tensions, strains, and risks that lead to conflict, war, and genocide.

It is my hope that you, the reader, finds this book to be informative and useful in helping make sense of the threats to our communities and societies as a consequence of a rapidly changing climate and environment. As I indicated earlier, this book examines the implications of climate change for creating and exacerbating a variety of risk factors that increase the likelihood of violent conflict, with a particular emphasis on genocide. In writing it, I have tried to avoid technical and theoretical jargon in order to make the reading as engaging and accessible as possible. I hope it appeals to a wide range of potential readers including experts and nonexperts alike. It is intended for anyone interested in climate change and its potential impacts on humanity.

I

MAKING SENSE OF CLIMATE CHANGE

In many ways the problem of human-induced climate change is unique: it is global, it will affect the planet for decades to centuries, and it is complex, imperfectly understood, and has the potential for truly dramatic consequences.

—Gavin Schmidt and Joshua Wolfe[1]

The climate has always been changing. On every timescale, since the Earth was first formed, its surface conditions have fluctuated. Past changes are etched on the landscape, have influenced the evolution of all life forms, and are a subtext of our economic and social history.

—William James Burroughs[2]

The more scientists learn about the natural climate revolutions woven into the fabric of the planet, the greater our awe about how supremely fickle is climate on earth.

—E. Kirsten Peters[3]

The winter of 2013–2014 was one for the record books in the United States and Canada. For much of December and January huge swaths of North America were locked into an Arctic freeze with many cities across the Midwest and East Coast experiencing record-breaking or near record-breaking low temperatures. O'Hare airport reached a low of –16 degrees Fahrenheit on January 6, while Detroit recorded a low temperature of –14 degrees Fahrenheit on the same day. At least forty-nine cities across the country recorded record lows that January.[4] The cold snap reached all the way down to Florida, Georgia, and Texas, states not generally experiencing cold winter temps. Up north, Lake Superior was almost entirely frozen over and as late as March around 95 percent of the lake was still covered in ice.[5] Not noted for a sense of humor or for the dramatic, the National Weather Service responded to the unusual cold by adopting the hashtag "Chiberia" for its reports from Chicago. This cold winter was not limited to North America. The United Kingdom reported the coldest March since 1962. One snowstorm in mid-March left portions of England and Northern Ireland under almost a foot of snow and killed thousands of newborn lambs.[6] The winter in Germany and Austria was similarly cold, while Russia, a bit farther to the east, experienced record snowfalls in Moscow. This extreme weather unleashed a veritable blizzard of commentaries about the weather from various journalists, media pundits, and others, with not a few calling into question the notion of climate change. For many, the logic seemed to be that if we have a cold winter, then how could the globe be warming up? One online post suggested that "for the first time in over a century Niagara Falls reportedly froze over in January 2014. Lake Michigan was over 90% ice covered this winter—another record. But despite the facts—the left continues to push their global warming junk science. And, if you disagree with them they want your [*sic*] jailed—for junk science."[7] Similar kinds of comments and reactions littered the airwaves and blogosphere. The evidence, however, continues to mount.

Louisiana suffered from catastrophic flooding in the summer of 2016 because of massive rains that pounded the state. At least thirteen people lost their lives and thousands of homes and businesses were destroyed or damaged.[8] Using new modeling techniques, climate change scientists found that such storms are at least 40 percent more likely because of climate change. Summarizing the research, one scientist asserted that "climate change played a very clear and quantifiable role."[9] And the list goes on. The summer of 2016 saw record heat waves in India and Pakistan that melted pavement and killed over a thousand people.[10] England experienced its hottest summer ever with London recording a high of 98 degrees Fahrenheit. From wildfires to drought to flooding, 2016 was a record-breaking year in so many categories. Yet, in some cases, the denial continues.[11] Commenting on such an irrational persistence, the economist Nicholas Stern asks the question, "How is it that, in the face of overwhelming scientific evidence, there are still some who would deny the dangers of climate change? Not surprisingly, the loudest voices are not scientific, and it is remarkable how many economists, lawyers, journalists and politicians set themselves up as experts on the science."[12] While political and economic forces often play important roles in fostering climate change denial, at other times it has more to do with a lack of understanding about what climate change actually is and how it works. Climate change is not something that can necessarily be seen from any particular weather event or from any one season. One winter or one storm doesn't necessarily mean much in terms of climate. Weather and climate are not the same and comments asserting that global warming can't be happening because of a cold winter, for example, reveal a fundamental misunderstanding of the difference between the two.

Weather and climate, although closely connected, are different phenomena. A thin membrane of gases, which we call the atmosphere, surrounds our planet. Extending upward for about a hundred miles, this thin, life-protecting film of insubstantial vapor is subject to various forces that set it in motion and create weather. Concentrated in the lowest and thickest layer of the atmosphere known as the troposphere, weather is largely a product of temperature gradients. Sunlight, or more precisely solar radiation, falls upon the earth and this energy provides heat. The surface of the planet, however, warms at different speeds depending on things such as the angle of the sun, whether the solar rays are falling on water or land, and cloud cover. Even the amount and type of vegetation can affect how quickly and how much a surface warms up from solar radiation. Essentially, this differential warming occurs because the albedo or reflectivity of the surface upon which the solar energy is falling can vary tremendously. Darker surfaces absorb more energy and consequently get warmer quickly, while light-colored surfaces tend to reflect more energy away and thus stay cooler for longer.[13] This is why snow and ice fields, which have a high albedo, can be in direct sunlight and still feel very cold. Most of the energy striking the snow and ice is simply reflected off, preventing the snow and ice from heating up very much. Forests, on the other hand, being much darker, have a lower albedo and consequently warm up more because they absorb more energy from the sun's rays. The same is true for urban centers that are largely asphalt and concrete.

When the sun sets due to the rotation of the planet, the cooling of the earth's surface also varies tremendously depending upon many of these same factors. Some surfaces retain heat longer than others, as anyone who has ever spent time in a desert city such as Phoenix, Arizona, for example, can tell you. Such cities are often many degrees warmer than the surrounding desert at night because all of the concrete and pavement retain heat much longer than the natural ground cover of the desert. This differential warming and cooling creates temperature gradients that cause air to move sideways and to rise and fall. Warm air tends to rise, while cold air, because it is denser, tends to sink. Throw in the effects of the earth's rotation, geographic features such as mountain ranges and bodies of water, and moisture in the atmosphere and you have weather. Weather, therefore, usually refers to things such as wind direction and speed, precipitation, cloudiness, humidity, and all the other factors that make up the atmospheric conditions in a given area. Weather, in other words, is what you experience out of doors on a daily basis. When you bundle up to ward off a cold wind or open an umbrella because of rain, you are reacting to weather.

Climate, on the other hand, refers to these daily conditions averaged over time. When you plan for certain weather conditions because of where you live and the time of year, then you are reacting to climate. Whereas weather is short term, climate is more long term. Because of this difference in timescale, weather can vary widely from hour to hour, day to day, and year to year, while climate tends to be relatively stable. When we talk about climate change we are looking at long-term trends and patterns, not short-term weather events. Climate is usually computed by measuring temperature, precipitation, air pressure, and wind over a period of years in order to generate a reasonable understanding of what can be expected in any given time and place. But climate is more than just a list of averages and statistics. Our understanding of climate encompasses a broader web of connections that includes the atmosphere, the lithosphere (land), the hydrosphere (water), the cryosphere (ice), and the biosphere (life). All of these different realms are connected and interact and influence each other in multiple and complex ways. So when we talk about climate change we are talking about a connected series of events that have impacts beyond the most immediate and directly affected aspect of climate. These connections are complex, multifaceted, and therefore often hard to understand, especially given the politically contested nature of the conversation surrounding climate change in certain segments of American society. Change in one sphere can have dramatic and often unforeseen ramifications in seemingly unrelated spheres. This is what makes predicting the effects of climate change so difficult. We simply don't know enough about all the various ways in which plants, animals, and weather interact, influence, and depend upon each other.

When discussing climate change, it is also important to understand that climate change has always been a part of this world. In one sense, it is nonsensical to talk about whether climate change is actually happening. There is no real debate. Climate has always been and will always be changing. Earth's climate has been seesawing back and forth between ice ages and warmer periods for billions of years. We tend to think of earth's climate as being stable and unchanging, probably because our time here on earth as a species has been so short. The existence of humanity can be encompassed within a tiny fraction of the most recent part of earth's history, which tends to give us a somewhat skewed perspective on the matter. It also doesn't help that any single person's lifespan is also quite short, which makes it difficult to see long-term changes. Humanity has actually lived through dramatic climate changes, including at least one ice age, but for the last ten thousand years the earth has been very warm and remarkably stable, at least in geologic terms. This current interglacial period, which scientists have termed the Holocene era, has allowed us to flourish and has contributed to the rise of human civilization, but these conditions certainly haven't been typical over the long course of earth's history. To really understand climate change, then, it is essential that we step back and take a longer-range perspective in order to understand that earth's climate has never been stable over a geologic timescale. So why does climate change?

The engine that drives earth's climate is, of course, the sun, or more specifically, the solar radiation that reaches the earth. Therefore, most natural sources of climate change relate in some way to the sun. Earth's climate is dependent upon the amount of solar energy received. It's actually quite simple, at least in principle. The more radiation we receive, the warmer the earth is. Over the course of the last few billion years, the amount reaching the earth has sometimes varied for a number of different reasons. We now know that solar output sometimes changes, a fact that seems to be reflected in the amount of sunspot activity. Sunspots, which look like dark spots on the surface of the sun, appear to be temporary phenomena that are linked with the intense magnetism generated by the sun and convection currents that circulate within the boiling nuclear cauldron of the sun's atmosphere. These sunspots appear to vary in a regular and predictable fashion with periods of greater sunspot activity being times of higher solar output.[14] Evidence suggests, for example, that a period of very low sunspot activity during the seventeenth century resulted in a temporary cooler climate that is now referred to as the "Little Ice Age."[15] During this period, glaciers in Europe advanced and overwhelmed high alpine villages, markets were held on the frozen Thames in winter, the Dutch famously skated on their ice-covered canals, and floods and storms ravaged the European continent with regularity.[16] For several centuries, the mean temperature of Europe was about 1.8 degrees cooler than it is now and many now believe that this climatic event was connected to sunspot activity. North America experienced exceptionally cold winters, while many African mountains were covered in snow year-round, and Timbuktu was flooded at least thirteen times by the Niger River, something that hadn't occurred before or since.[17] Other sources of climate change, although linked with the sun, have more to do with the movement of our planet.

The earth's axis has a tilt or obliquity, which over the course of forty-one thousand years or so cycles back and forth between 22 degrees and 24.5 degrees. This changes the angle at which the sun's rays hit the planet, and since it is the tilt of the earth that gives us seasons, the greater the tilt the bigger the difference between summer and winter. During eras with more tilt, winters in the Northern Hemisphere become more extreme, which allows vast ice sheets to form, thereby increasing the reflectivity of the earth and helping to trigger ice ages.[18] The increase in ice cover, in other words, amplifies the effect of the change in earth's tilt and thus helps create the conditions for an ice age. Furthermore, over the course of around a hundred thousand years, the earth's orbit also changes from an elliptical or eccentric path to a more circular one and this also dramatically affects earth's climate since it changes the distance the earth is from the sun. Our orbit is mostly circular at present, which means that there is only a 6 percent difference in the amount of radiation reaching the earth between winter and summer, but when the orbit is more elliptical that difference can range from 20 to 30 percent.[19] These variations tend to bring about protracted changes lasting many thousands of years or even longer. By themselves, these orbital and axial changes do not necessarily bring about ice ages. Rather, their effects are augmented by feedback loops within the climatic systems that amplify and accelerate these orbital changes. Once the earth begins cooling because of the earth's tilt and/or orbital path, for example, ice sheets can develop at the poles and spread outward. This increase in surface ice raises the albedo of the planet, thus amplifying the cooling effect. Keep in mind that these changes are ones that occur over the course of thousands of years and estimates suggest that it generally takes about forty thousand years for the climate to be pushed into an ice age.[20]

We now know that during the approximately 4.6 billion years that this planet has existed, there appear to have been four major ice ages lasting millions of years.[21] In fact, the second one was probably so severe that ice completely covered the surface of the planet creating what some have called "snowball earth."[22] But most were not quite as bad. It might be hard to believe, but the warm period we are currently living through has been more the exception than the rule. Simply put, the history of this planet has been largely dominated by ice and cold. The most recent ice age ended about ten thousand to fifteen thousand years ago and we've been living in the warm aftermath ever since. These ice ages and the interglacial periods in between, however, have not been completely uniform. During the last ice age, for example, rapid climatic shifts frequently occurred, shifting the global thermostat back and forth between warm periods and cold ones.[23] The most recent era is known as the Holocene and has been characterized by overall warmer and milder global conditions. I say overall, because like the preceding ice ages, the Holocene has also been characterized by sharp and sudden changes in climate. Much shorter in duration and relatively shallow in terms of overall impact, these climatic oscillations are believed to be partially caused by a complex and little understood interplay between the oceans and the atmosphere. Europe, for example, has endured a great deal of climatic variation since the last ice age, from the Medieval Warm Period, which allowed the Norse to settle Greenland, vineyards to flourish in England, and the cultivation of marginal lands high in the mountains of Europe and Scotland, to the Little Ice Age, which reversed many of these trends and created widespread havoc throughout Europe. These oscillations have periodically reversed climatic trends in various regions and have produced short sharp shocks for the human communities that have had to cope and endure with them and, noteworthy for our discussion, these fluctuations have often been correlated with social dislocation, societal unrest and collapse, conflict, war, and genocide, issues to which we will be returning in subsequent chapters.

In addition to the celestial sources of climate change discussed above, there are other more terrestrial ones as well. Volcanic eruptions, for example, have sometimes thrown up enough ash into the atmosphere to affect the earth's climate, at least temporarily. Normally, about 30 percent of the sun's rays are reflected back into space by clouds and other particulates in the atmosphere, and by light-colored surface areas of snow, ice, and desert sand.[24] Eruptions can increase the reflectivity of the atmosphere markedly. The volcanic ejecta can circulate up in the atmosphere for a long period of time, blocking sunlight from reaching the surface and effectively cooling the planet. The eruptions of Tambora in 1815, Krakatoa in 1883, and Pinatubo in 1991 spewed enough dust into the atmosphere so that global temperatures dropped dramatically for several years afterward until the particulates had descended and sunlight was able to warm the earth again. Even earlier, approximately seventy-three thousand years ago, a volcano erupted, but it was not just any volcano. It was Mount Toba in what is now Sumatra and it was one of the largest eruptions ever. The eruption spewed ash twenty miles into the atmosphere and created lava flows that covered over seven thousand square miles.[25] Ultimately, it created a crater lake that is still in existence. It is sixty-two miles long, eighteen miles wide, and over a thousand feet deep. Early humanity almost became extinct from both the short-term and long-term consequences of the eruption. Those who didn't die from the suffocating ash died from the lack of plant and animal life, which also perished in the aftermath of the eruption. Much of the surface of the earth was covered in ash, up to nine feet deep in some areas, while the global climate experienced a dramatic cooling effect lasting almost two thousand years. The early humans that survived the immediate eruption and subsequent fallout had to cope with a much colder environment, one without the same abundance of plant and animal life and very few endured that transition. The result was a genetic bottleneck that is still written into our DNA.[26] Early humanity almost never had a chance to get a foothold on this world due to volcanic activity.

In addition to these natural sources of climate change, there is also anthropogenic or human-induced change and, for all practical purposes, this is what people mean when they talk about climate change in recent years. Often referred to as global warming, this term refers to the heating up of the earth's climate due to human activity. By this point, the evidence overwhelmingly indicates that the earth is warming up and that a big part of the reason has to do with our consumption and burning of fossil fuels that create greenhouse gases. These emissions serve to insulate our atmosphere and result in warmer average temperatures around the world.[27] So what are these greenhouse gases and how do they work?

The main greenhouse gases are carbon dioxide, which is produced by natural sources such as decomposition and volcanic activity, but since the Industrial Revolution have been increasingly generated by human sources such as the burning of fossil fuels and deforestation; methane, which comes naturally from wetlands and from human activities such as agriculture; nitrous oxide and ozone, which also come from agriculture and burning fossil fuels; and halocarbons that are derived from natural sources such as forest fires but also from refrigerants and other industrial materials that may be released. Once in the atmosphere, these gases act like a blanket over the surface of the earth. Essentially, the atmosphere allows much of the sun's rays to penetrate to the surface of the planet where they warm the earth and the oceans. This heat is then radiated back out, but since it is at a lower energy level it is trapped close to the surface by the greenhouse gases. Of course, earth has always had greenhouse gases in the atmosphere, which has made the earth warm enough for life to develop and survive. Without greenhouse gases, the surface of our planet would be around −.4 degrees Fahrenheit and the planet would exist in an unending deep freeze, lifeless and barren. Some greenhouse gases, then, are essential for life to exist on earth. Without them our planet might be like Mars, which has a very thin atmosphere but nowhere near enough to trap solar heat or support life. Consequently, the average temperature on Mars is about −80 degrees Fahrenheit.[28] The situation on earth now, however, is that humanity is adding huge amounts of greenhouse gases

to the atmosphere in a very short period of time and causing the opposite effect.

While human beings have always affected the climate, until the last
few hundred years, our impact has always been small enough to be
offset by natural forces and processes. Beginning in the late eighteenth
century, however, that reality changed. At that time, civilization went
through a significant transition that fundamentally altered the way humans live, so much so that some have suggested that this modern time
period should be referred to as the Anthropocene—the Human era—
instead of the Holocene because of humanity's impact on the world.[29]
The Industrial Revolution was largely responsible for this transition as it
heralded the advent of material abundance and wealth that are unprecedented in human history. Unfortunately, it also depended upon the
burning of fossil fuels. The Industrial Revolution was quite literally
built upon the substitution of manual labor and simple cottage indus-
tries with machine-based production and manufacturing. Essentially, it
signaled a transition from small-scale to large-scale and from handmade
to machine-made goods and services. Largely beginning in England,
then spreading through Europe and from there around the world, the
Industrial Revolution relied upon machinery to produce goods much
more quickly and efficiently than people had previously been able to
do. The machines of industry, however, have voracious appetites and
require constant feeding. Initially, wood was the fuel of choice, but it
quickly became apparent that England and Europe's forests would soon
be exhausted, so industry turned to coal to power the steam engines of
transportation and commerce.

Coal, the most plentiful of fossil fuels, started out as plant matter in
countless swamps around the world during eras such as the Eocene and
Carboniferous when swamps covered much of the earth. Over time, as
vegetation died and sank to the bottom, thick layers of vegetative matter
formed because a lack of oxygen at the bottom of these swamps im-
peded decomposition. After enough time had passed, and enough sand
and silt were deposited on top of the thick layers of dead plants, the
pressure and the heat would mount as this material was buried deeper
and deeper. After millions of years of heat and pressure, this organic
material was transformed into coal.[30] Even to the present day, coal
remains the single most widely used type of fossil fuel around the world
because it is cheap and abundant. Because of this, coal plants continue

to be built in large numbers, especially in developing economies such as China and India—a trend that is expected to continue well into this new century. In November 2015, for example, the Chinese government announced that it had issued permits for the construction of 155 brand-new coal plants.[31] Oil was the next fossil fuel to be exploited. Whereas coal was formed from dead swamp matter, oil is the product of dead phytoplankton that sank down to the ocean depths where oxygen is in short supply, and where the same geologic forces that created coal served to transform the accumulated corpses of the single-celled plants known as phytoplankton into the viscous fluid that is so sought after today because it facilitates our addiction to the automobile. The third fossil fuel that appears to be increasingly relied upon, especially in the coming years if projections are correct, is natural gas, which is largely composed of methane.[32]

These three fossil fuels—coal, oil, and gas—have become the fuels of choice because they have been cheap, relatively abundant, and fairly easy to get and provide a great deal of energy when burned. Unfortunately, these fuels also contain high levels of carbon and when burned for energy in homes, factories, and automobiles release large amounts of carbon dioxide as a waste product. Prior to the Industrial Revolution windmills and watermills usually provided power, and people burned wood and/or peat for heat. Both wood and peat also released carbon dioxide, but because the population was small, the carbon output was small enough so that the cumulative effect on the climate was negligible and offset by natural processes that reabsorb carbon dioxide and take it out of the atmosphere. Since the Industrial Revolution, however, the amount of carbon dioxide pumped into the atmosphere has increased exponentially. At first it was primarily from factories, then increasingly refineries and power plants, and now automobiles, farm machinery, and the like. Today, the seven billion human beings on planet earth use, on average, four times as much energy as people did a hundred years ago, and as a result, the use of fossil fuels has increased by a factor of sixteen.[33] From the era before industrialization to the present day, atmospheric carbon dioxide has increased from 280 parts per million (ppm) to 384 ppm, while methane levels have grown from 700 parts per billion (ppb) to 1,857 ppb.[34] Other greenhouse gases have also increased substantially. This increase in atmospheric levels of carbon dioxide is the greatest it has been for the past 650,000 years and has resulted in a mean global temperature increase of around 1.3 degrees Fahrenheit since the onset of the Industrial Revolution.[35]

Fossil fuels are not the only source of greenhouse gases, but they are the primary cause of global warming. Much of the industrialized world is waking up to threats posed by climate change and some countries have begun taking steps to curb their emissions through more efficient use of fuel and a greater reliance on alternative fuels. Unfortunately, much of the damage is already done. The atmospheric life span of carbon dioxide is anywhere from 5 to 200 years, for nitrous oxide it is 114 years, for carbon tetrachloride it is 85 years, and for methane it is 12 years. This means that even if the emission of greenhouse gases were somehow to miraculously end today, the impact of these contaminants on the world's climate would still be with us for a long time to come. If this isn't alarming enough, in much of the developing world, emissions have increased dramatically in recent years. In their zeal to reap the benefits of a market economy and consumerism, China and India, for example, have worked hard to modernize their economies, which has translated into more coal-based power plants and a rising standard of living for their citizens. This has resulted in correspondingly higher energy consumption as Indians and Chinese citizens buy more automobiles, build bigger homes, and buy more food and consumer goods. This pattern is being replicated in many places around the world. Between 1990 and 2000, for example, the world transportation sector alone (cars, trucks, shipping, and airplanes) produced 36 percent more greenhouse gases. China is motorizing its society faster than any other in history. The city of Beijing, for example, had 2,300 cars on the streets in 1949, which increased to 2,000,000 in 2003, and is climbing by about 1,000 new vehicles a day. With similar patterns occurring across the country, and a total population in China of 1,324,655,000, such a growth in automobile ownership and usage has a sizable impact on greenhouse gases.[36] But it is not just countries with large populations that are affecting the ozone. Americans, who compose only about 4 percent of the world population, consume just under 20 percent of the yearly expenditure of energy.[37]

In the past, the carbon emissions produced by humanity and by various natural sources were readily absorbed by what are termed "carbon sinks." One of the most potent of these sinks is the world's oceans, but as the world warms, the oceans will also warm, cutting down on their ability to absorb carbon and other greenhouse gases. Since 1955, for example, over 90 percent of the extra heat retained on earth due to greenhouse gasses has been absorbed by the world's oceans.[38] In fact, scientists have already documented large-scale releases of methane into the atmosphere from the Arctic Ocean, as well as from thawing permafrost.[39] Climate experts fear that this kind of feedback loop can greatly accelerate the impact of global warming. The cutting down of many forests around the world has also had a similar negative impact. Because of the relentless demand for wood and for cleared land, deforestation, especially in places such as Brazil, Indonesia, and Madagascar, has resulted in half of the world's forests being destroyed at a rate of around one acre per second.[40] Cutting down trees has two main climate impacts. First, trees absorb carbon dioxide from the atmosphere and store it in their biomass, and second, deforestation through burning releases the trapped carbon back into the atmosphere. Eliminating or weakening the natural mechanisms for trapping carbon, therefore, also serves to increase the effects of anthropogenic climate change. So what does all this mean? What are the consequences of the changes that humanity has unwittingly set into motion?

In 2014, the Intergovernmental Panel on Climate Change (IPCC) released its latest report, asserting that from 1880 to 2012 the combined average land and ocean surface temperature increased .85 degrees Celsius or around 1.3 degrees Fahrenheit.[41] While there have been yearly and even decadal variations, and while not every region of the world has experienced the same increase, the average temperature of our planet has increased significantly. In January of 2016, NASA and the US National Oceanic and Atmospheric Administration (NOAA) announced that 2015 was the warmest year on record since 1880, the year that record keeping began.[42] Moreover, this record was a whopping 20 percent higher than the previous warmest year, which was 2014. This is the reality of climate change, and 2016 continued the trend by becoming the hottest year on record. How does this warmer world play out in detail? Unfortunately, there are far too many consequences of climate change to be able to discuss all of them. To further compound the difficulties, the consequences will play out differently depending upon the region of the world and the steps taken to ameliorate the effects by individual nations. Climate change will not affect the world equally.

That being said, we can take a look at a few of the most obvious and compelling consequences.

One effect of a warmer world that is already quite visible is sea level rise. Data reveal that sea levels have increased about eight inches since 1880 and the rate of rise is accelerating.[43] The IPCC reports that between 1901 and 2010, the mean rate of sea level rise was 1.7 millimeters per year. This increased to 2 millimeters per year between 1971 and 2010, and 3.2 millimeters per year between 1993 and 2010.[44] While prediction is always an uncertain game, some estimates suggest an increase of 10 to 20 centimeters by 2030, and an increase of 1 meter by the end of the twenty-first century.[45] As with other effects this won't necessarily be uniformly distributed. The 1993–2005 satellite data, for example, indicate that during that time period the western Pacific increased five times as much as the global average, while the eastern Pacific actually decreased slightly.[46] But, on average, sea levels will increase fairly significantly because of a number of different processes that contribute to these increasing levels.

First, there is a thermal expansion effect. As the oceans warm, their volume increases and the water takes up more space. This is the same principle we see with thermometers, which depend on liquid expanding and rising within sealed glass tubes when the temperature rises. Second, much of the earth's water is locked up as ice and as the earth warms this ice will melt and contribute to sea level rise. In the north, the Arctic Ocean is a body of water roughly 2,800 miles in diameter that spends much of the year under a layer of ice anywhere from three to six feet thick. By the end of the twentieth century, the sea ice of the Arctic had decreased by 25 percent and was significantly less thick than it had been at midcentury.[47] In fact, measurements indicate that the ice thickness is only about 60 percent of what it was forty years ago.[48] It is important to note that the melting of Arctic sea ice does not contribute directly to sea level rise, since the ice is already afloat and cannot displace any extra sea water as it melts. Instead, it contributes to sea level rise indirectly as the loss of sea ice changes the earth's albedo in a way that increases warming and hastens the melting of continental ice in glaciers, most notably the Greenland ice sheet. This particular mass of ice is so large that it is up to two miles thick at its maximum and so heavy that it actually depresses the rock surface of Greenland by about three thousand feet.[49] If all the ice in Greenland were to melt, sea levels would rise by about seven meters or twenty-three feet.[50] While initial projections suggested that these impacts weren't in the cards for the near future, recent research has identified some worrying possibilities. In 2004, for example, scientists found that Greenland's glaciers were melting ten times more rapidly than originally believed.[51] Then there's Antarctica.

Evidence suggests that Antarctica is warming more rapidly than other parts of the globe with potentially very dramatic consequences since 90 percent of the world's glacial ice is contained in Antarctica.[52] Antarctica is a continent that is hemmed in by vast sheets of sea ice that are beginning to melt with big chunks periodically breaking off and floating away. Additionally, the remaining ice shelves are thinning due to the water they rest in becoming warmer. This process has also accelerated the loss of the continental ice since the ice sheets serve as a barrier for the glaciers on land. In other words, glaciers flow downhill and the flow of glaciers to the sea has always been restricted by the amount of ice at the glacier's mouth. Sea ice acts as a plug on glacial flow. When removed or weakened, the rate at which glaciers flow increases and this is what appears to be happening in Antarctica.[53] Since 2002, the southern continent has been losing over twenty-four cubic miles of ice per year. Most of this loss is coming from West Antarctica. If this ice sheet is completely lost it could raise sea levels by sixteen to twenty-three feet.[54] If all of Antarctica lost its sea ice, projections suggest a sea level rise close to two hundred feet, a truly apocalyptic catastrophe, which at this point some suggest might eventually be inevitable.[55] In the summer of 2016, James Hansen, one of the most influential climate change scientists and the man who first brought the issue to the attention of the public in 1988, revealed that a newly identified feedback mechanism of the Antarctic coast strongly indicates that sea level rise could occur ten times quicker than earlier estimates suggested.[56] By 2065, he warned, a sea level rise of ten feet is quite possible with catastrophic effects for coastal regions and communities, not to mention island nations.

In summary, because of the loss of glacial ice worldwide, and because of thermal expansion, it is clear that sea levels around the world will be increasing for the foreseeable future. Perhaps the most obvious effect of sea level rise is that a great deal of land will disappear under the waves. Coastal habitats, river deltas, and islands will be inundated with severe consequences for the ecosystems of those affected areas.

Given that about 40 percent of the world's population lives within sixty miles of the coast and around a hundred million people live less than four feet above sea level,[57] it is clear that sea level rise represents not just an ecological catastrophe, but a social, economic, and political one as well. The massive influx of cold freshwater into the world's oceans also will likely disrupt and alter weather patterns around the world.

Traditional patterns of weather are largely created by a complicated interplay between the atmosphere and ocean temperatures and currents. The oceans of the world are subject to various currents that circulate water around the world. Known as the ocean conveyer belt or as the thermohaline circulation, these oceanic rivers transport water and help to regulate the temperature of the planet. Perhaps the most well-known ocean conveyer belt is the Gulf Stream, which is a surface current that brings warm water from the Caribbean into the North Atlantic.[58] Five hundred times larger than the Amazon river, the Gulf Stream is formed in the Caribbean where the sun heats the water that is pushed by surface winds up the eastern seaboard of North America and then east across the Atlantic past Ireland and Iceland to finally end in the Arctic waters between Greenland and Norway. As it travels it also cools, which increases its density until it finally downwells into the depths to become part of the North Atlantic Deep Water current, which is a flow of cold and deep water that courses south into the South Atlantic and then joins with currents around Antarctica and from there into the Indian and Pacific Oceans.

This global conveyer belt serves as a heat regulator distributing warmth around the globe and helps keep the earth relatively warm. The Gulf Stream, for example, keeps Iceland, Ireland, Great Britain, Scandinavia, and Europe much more temperate than one would expect given their high latitudes. Compare the climate in Ireland and Scotland, for example, with that of Labrador and Alaska, which rest at comparable latitudes but do not benefit from warm ocean currents. Warmed by the Gulf Stream, Ireland and Scotland experience much milder winters than do Labrador and Alaska.[59] It seems that the Gulf Stream has even been powerful enough to prevent the earth from entering another ice age in the past.[60] The melting of the Greenland Ice Pack will result in large amounts of freshwater mixing with the water of the Gulf Stream somewhere south of Greenland. Because freshwater is more buoyant than saltwater, this influx may prevent the downwelling that helps drive the Gulf Stream and consequently weaken or shut down this ocean current with significant consequences for the relatively balmy European climate.[61] This has actually happened in the past. The Younger Dryas period, a global cold snap that happened around thirteen thousand years ago, occurred after Lake Agassiz broke its banks and released a huge amount of freshwater into the North Atlantic. Located across much of what is now the Canadian state of Manitoba and the upper American Midwest, this massive lake had been formed by the melting of the North American ice sheets and when it spilled over into the North Atlantic, the enormous influx of freshwater into the Gulf Stream effectively shut down this oceanic conveyor belt and northern Europe quickly plunged into arctic conditions.[62] It took over a thousand years for the Gulf Stream to start up again, bringing a more temperate climate to northern Europe once again.[63]

The bottom line is that a warming earth will significantly disrupt global patterns of weather. As the earth warms, predictions suggest that weather patterns will adjust with some regions seeing increased rainfall and moisture, while other areas will experience increased and prolonged droughts. Predictions suggest, for example, that Northern Africa will have significantly less rainfall in the decades to come, while Eastern Africa will experience more precipitation during its rainy season. Central America and northern South America will probably experience less rainfall, while large parts of western South America may see more rain. The northeastern United States will in all likelihood see more rainfall, while the southwest and south central states will experience drier conditions. Potable water, agricultural productivity, and population distribution will all be dramatically impacted by these changes in traditional patterns of precipitation. Many agricultural areas around the world, for example, depend heavily upon irrigation and if traditional sources of water in those regions are diminished or lost, those areas will see lower yields and crop loss. This means that many nations will see their ability to grow crops and feed their populations weakened as traditional rain patterns change, rivers, lakes, and aquifers diminish or run dry, and crops succumb to the combined effects of less precipitation and higher average temperatures. Many crops have a very limited temperature range and are already close to their maximum.[64] In fact, research suggests that a 4 degrees Celsius rise in average temperature would render half of the world's agricultural farmland useless for growing crops.[65]

The ability of a nation to feed its population can be surprisingly fragile since humans depend on only three staple crops (wheat, corn, and rice) for around 90 percent of all caloric intake.[66]

Another outcome of a warmer world will be more frequent and severe weather events, such as storms, tornados, and hurricanes.[67] In fact, as I noted above, some observers believe that this reality is upon us and suggest that once-uncommon weather events have become much more prevalent, including extreme heat waves, longer and deeper droughts, heavy rains and flooding, wildfires, and more frequent hurricanes with higher wind speeds and storm surges.[68] Scientists, for example, argue that the loss of polar ice makes extreme winters like that experienced by much of North America in 2014 and 2015, briefly discussed at the beginning of this chapter, more likely.[69] Keep in mind that we are only now at the leading edge of these changes. These will not only continue, but will in all likelihood accelerate.

Coastal regions will experience sea level rise from the melting of polar and glacial ice. When combined with the thermal expansion of seawater, many seaside communities and areas will be at high risk for outright inundation or at the very least subjected to higher tides and storm surges periodically flooding coastal land. In November 2013, Typhoon Haiyan, one of the strongest tropical cyclones ever recorded, made landfall in the Philippines and Vietnam and killed over six thousand people and resulted in at least $13 billion in damages. These once-rare events are expected to increase in frequency. The impact of such weather, however, is not limited to the easily seen physical devastation wrought by such events. Supplies of potable water, for example, will also be diminished in shoreline areas because of the increasing salinity of rivers, bays, and coastal groundwater sources while loss of coastal fisheries due to warmer water will negatively impact those who depend on fishing for their livelihoods and for food.[70] Warmer oceans will not just breed more extreme storms but will also further stress coral reefs already damaged by pollution and rising waters. In turn, this will further threaten marine life that depend upon coral reefs, as well as endangering the shorelines protected by reefs. Moreover, as the oceans warm they become more acidic, which will pose significant problems for ocean biodiversity and those communities that depend on fish and other seafood to sustain themselves.

On land, entire ecosystems will shift in response to changing climatic conditions. Some ecological zones will move geographically as changing temperatures and rainfall patterns render formerly life-supporting areas inhospitable for particular plants and animals. On the other hand, other previously hostile zones will become more welcoming. In those locations, however, where there is no available land to allow ecologies to shift or migrate into more hospitable locations, entire ecosystems will disappear and the flora and fauna dependent upon them will go extinct, at least in that location. The biologist Anthony Barnosky puts it succinctly when he writes:

> Today, it is not an option, for a couple of very important reasons, for many species to alter their range to follow their needed climate. First, climate change is racing faster than it ever has during the evolution of living species and ecosystems—many species simply aren't biologically capable of adjusting their geographic range at the speed they would need to in order to survive. . . . With cities, towns, large-scale agriculture, roads, and other impediments, we have fragmented the natural geographic ranges of many species and at the same time thrown barriers in the path that would otherwise make it possible for them to move freely around the Earth's surface, even if they could keep up with climate change. As a result, whole communities and ecosystems may fail to operate as they have evolved to do over thousands, even millions, of years. Under such conditions, species may not be able to adapt through natural selection as they have done in the past because the speed of climate change is simply too fast for evolution to keep up.[71]

We know that many animal species are already threatened and under stress because of loss of habitat through development. This has been going on for a long while now. Plant and animal species will have to adapt or die. It's that simple. One recent study found that almost half of all mammals and almost a third of all birds that are on the International Union for Conservation Nature's Red List of Threatened Species have already been negatively impacted by climate change.[72] The authors of the study conclude with the following recommendation:

> Our results suggest that the impact of climate change on mammals and birds in the recent past is currently greatly underappreciated: Large numbers of threatened species have already been impacted in

at least part of their range. Given that scientific efforts in this field have largely focused on predicting the impact of future climate change on species and ecosystems, we recommend that research and conservation efforts give greater attention to the "here and now" of climate change impacts on life on Earth.[73]

In other words, far more vulnerable species than previously believed are already struggling to cope with climate change and this trend is only expected to increase and intensify. This scenario has played out before. At the end of the last ice age, a great number of large animals, often referred to as megafauna, went extinct. Sabre-toothed-tigers, giant deer, mammoths, wooly rhinos, and the great cave bear, for example, all vanished from the American landscape when the climate changed and these species were unable to adjust enough to survive. While some argue that early humans hunted them into extinction, many others suggest that a changing world is what did them in.[74]

Since life first arose almost four billion years ago, there have been five mass extinctions. The first was the Ordovician-Silurian that took place around 450 million years ago.[75] In all likelihood, the cause was an ice age that contributed to a change in the chemistry of the oceans. Extreme glaciation also meant that sea levels dropped dramatically and thereby destroyed the habitats of many marine life-forms. Most life on earth at that time existed in the water and this event wiped out around 85 percent of sea life. Next was the Late Devonian extinction that occurred around 400 million years ago, and this was followed by the Permian extinction about 250 million years ago. This latter one is sometimes called "The Great Dying," since around 96 percent of all species on earth went extinct. Possible explanations for this event include asteroids hitting the earth, massive eruptions of methane, variations in oxygen levels, changes in sea levels, or some combination of these catastrophes. Scientists have found that at the same time as this extinction, a huge amount of carbon was released into the atmosphere. This greenhouse gas caused a dramatic rise in atmospheric temperatures and the oceans warmed by as much as 18 degrees Fahrenheit. Because of this increase, the chemistry of the oceans changed and they became more acidic. Warmer water also carries much less oxygen, which meant that much sea life simply suffocated.[76] Next up was the Triassic-Jurassic mass extinction around 200 million years ago, followed by the Cretaceous-Tertiary about 65 million years ago. These last two were also likely caused by a variety of forces including climate change, volcanic eruptions, and asteroid impacts. It was the Cretaceous-Tertiary, often referred to as the KT extinction, which is arguably the most famous mass extinction because it was the one that saw the end of the dinosaurs. Current thinking suggests that the climate was already changing because of volcanic activity. Already stressed by a changing environment, the dinosaurs were pushed over the edge into extinction when a large asteroid or comet impacted the earth near the Yucatán Peninsula in Mexico.

These mass extinction events are important to point out since not only do they reveal how changes in the global climate have impacted life and biodiversity on this planet in the distant past, but also there is increasing evidence that we are on the leading edge of a sixth mass extinction event, this one caused by human beings.[77] Humanity has had a huge impact on plants and animals by destroying natural habitats through development, logging, damming of rivers, urban sprawl, and similar means. More indirectly, we are impacting the natural world through anthropogenic climate change. Many species are already reacting to a warming world. For example, migratory birds are changing age-old patterns, and plant species are trending farther north and/or higher into the mountains and blooming sooner than in the past. For the first time in centuries, anopheles mosquitos have been found north of the Alps, while butterflies are moving ever closer to the poles.[78] Tropical fish are appearing in the North Atlantic and the Mediterranean, while cod, a cold-water fish, have been trending farther north than they have been seen since the Medieval Warm Period.[79] These adaptations and changes are occurring all over the world as plant and animal species struggle to adapt to a changing world, some more successfully than others. Between 1970 and 2010, the World Wildlife Fund's Living Planet Index calculates that animal populations have been halved and that many bird species are particularly at risk because of a loss of suitable habitats.[80] While we often focus on the natural world and the environmental consequences of climate change, we also need to remember another species affected by climate change and that is us: humanity. Making this point, the social scientist Harald Welzer finds it astonishing that nearly all academic studies, models and prognoses regarding the phenomena and consequences of climate change have been in the *natural sciences*. In the social and cultural sciences, it is exactly as if such things as social breakdown, resource conflict, mass migration, safety threats, widespread fears, radicalization and militarized or violence-governed economies did not belong to their sphere of competence.[81]

This is precisely the focus of this book. To address the possible ways in which climate-induced stressors will impact societies and increase the risks for conflict, especially the more extreme forms of collective violence, most notably war and genocide. We, as a species, must respond to these challenges or as the anthropologist Richard Leakey pithily warns, "*Homo sapiens* might not only be the agent of the sixth extinction, but also risks being one of its victims."[82]

Climate change is as much a human and social issue as it is an environmental one and will present tremendous trials and difficulties for societies, nations, and communities. But this isn't a new issue. It's always been this way. We are only now realizing the extent to which perturbations in the earth's climate have shaped and altered the chronicle of human life in this world. Climate change has always helped shape human affairs. But this is a relatively new awareness. Traditionally, historians tracing the outlines of the past have largely focused on stories of wars, kings, and queens. When portraying the rise and fall of empires and civilizations, the usual story has been dominated by political maneuverings, the personalities of famous historical figures, and victories and defeats earned or lost on the battlefield. In recent years, however, this story line has been changing and evolving as we learn more and more about the role that climate has played in past times. Throughout history, the impersonal forces of nature have been constant companions to human existence and even the mightiest of empires has been profoundly influenced by the workings of the global climate machine. We tend to think of the earth's climate as being stable and unvarying throughout humanity's time upon this planet, but the reality is that it has changed frequently, sometimes dramatically, and those changes have spectacularly affected the course of human history from the very beginning.

Some have suggested that the evolution of early hominids into modern humans was shaped and influenced by geologic and climatic changes that helped them develop and also propelled them out of Africa to spread around the world.[83] Much of this migration, lasting thousands of years, was enabled by changes in climatic conditions. The settling of the Americas, for example, was made possible by the last ice age. The Bering Strait separating Alaska from Siberia is a shallow body of water, and during one of the last periods of heavy glaciation the ocean was around three hundred feet lower than it is today and created Beringia, a stretch of land connecting Asia and North America, sometimes referred to as the Bering Land Bridge.[84] In other words, the settling of the Americas was enabled by the climatic shifts that provided a passage for the people who would become the natives of the Americas. In more recent times, climate change helped create conditions conducive to the rise of early civilizations, but subsequent changes also sometimes helped bring about their collapse. Warm and stable conditions during the New Stone and Bronze Ages contributed to the development of complex societies in the Mediterranean, Mesopotamia, India, and China, but when these conditions ended, those civilizations often suffered as a result. The Indus or Harappan civilization, for example, sprang up during a time of plentiful and reliable rainfall during the Bronze Age along what is now the India and Pakistan border. It was a sophisticated society with well-developed systems of trade, metallurgy, irrigation and agriculture, and writing and the arts.[85] At its height, it was larger than the contemporaneous and more well-known civilizations of ancient Egypt and Mesopotamia combined. Its abrupt demise appears to have been largely brought about by a deteriorating climate that shifted monsoon rains away and caused local rivers to dry up, dramatically reducing the ability of Harappan society to feed and maintain itself.[86]

During the Middle Ages the climate pendulum swung back and forth several times with important consequences for early modern civilizations. For around five hundred years, from about 800 CE to 1300 CE, the world experienced what is often referred to as the Medieval Warm Period, an era marked by unusually warm and stable weather.[87] Warmer summers, milder winters, and a longer growing period meant growing populations in Europe as crops became more dependable, fewer people died from disease and famine, and more and more land was put under the plow. In this era, farms sprang up at higher elevations in the mountains, including valleys in the Alps that had previously been covered by glaciers. Crops grew as far north as Scotland and northern Norway. None of these areas have seen cultivation since; climatic conditions only permitted agriculture during this warm period. In Scandinavia, the increase in population density and a lack of flat and arable land combined with calmer seas resulted in the Norse surging out of their homeland and discovering and settling Iceland and Greenland, not to mention raiding and plundering coastal communities up and down Europe and into the Mediterranean.

Not every area benefited from the Medieval Warm Period. There were climate losers as well as winners. Long-term droughts devastated societies throughout Northern and Eastern Africa, the American Southwest, and Central and South America. More frequent and deeper droughts in the steppes of Central Asia resulted in the Mongol tribes unifying under the able leadership of Genghis Khan and embarking on the creation of a vast empire. Genghis Khan and his successors succeeded in conquering China, most of Asia, parts of the Middle East, Russia, the Ukraine, Poland, Hungary, and portions of Austria. The Medieval Warm Period was followed by an era known as the Little Ice Age, which as the name implies, meant a return of colder and more unsettled climatic conditions with predictable results for societies around the world. In Europe, this meant excessive rains, extremely long and cold winters, loss of livestock to epidemics, and crop failures. These conditions led to widespread famine, social disorder, and endemic warfare.[88]

In short, climate change has always been a part of human existence on this planet and the coming years will be no different. Nations and communities all around the world will be faced with changes that they must confront. These challenges will vary depending on localized conditions, and importantly, not every nation or community will have the same ability to respond effectively or even adequately. Much will depend upon the choices made by political leaders, social institutions, and communities as they struggle to cope with the myriad difficulties brought about by a warming earth. The strategies that local, regional, and national social and political leaders rely on will also depend upon the available resources within any particular nation or region. Some policies and approaches will be much more successful than others. The important thing to remember, however, is that climate change is not just about one or two short-lived challenges, but rather that these events will accumulate over time and may very well erode the ability of any political and social system to cope. Political scientist James R. Lee summarizes this reality well when he writes:

> People can survive aberrant, short-term climate change through exploitation of saved resources, but this strategy has temporal limits. The issue is not one of surviving a particularly fierce rain or a harsh winter, but the accumulation of many rain events and many harsh winters. Human society is capable of enduring events and seasons, but as these events and seasons accumulate over many years or even decades, accumulated wealth begins to draw down and eventually dissipates. Without renewal of society's wealth, human health and well-being decline, and over time the society itself may collapse.[89]

This is where the real danger lies. Given what we know about human history and the ways in which populations and their leaders can sometimes react to stressors, it is not unthinkable that in some places, eruptions of communal, religious, and sectarian violence, civil and international wars, and genocide will occur. In some cases, the violence will be instrumental in the sense that it will be seen as a means to an end, such as when a nation wages war to protect or acquire vital and needed resources, while in other cases, it may be expressive and reflect emotional responses to the fear, anxiety, and scapegoating engendered by the social, political, economic, and religious changes and insecurities brought about by climate change. Very often such conflict is built upon preexisting tensions and prejudice. To better understand the ways in which climate change can foster violent solutions and reactions, we need to have a better understanding of the various forms of collective and communal violence that may be carried out and it is to these that we now turn the discussion.

2

ON THE ORIGINS OF VIOLENT CONFLICT

War and the Genocidal Impulse

All across the planet, extreme weather and water scarcity now inflame and escalate existing social conflicts.

—Christian Parenti[1]

Climate has helped shape civilization, but not by being benign.

—Brian Fagan[2]

But, whether wars in the twenty-first century are directly or indirectly due to climate change, violence has a great future ahead of it. We shall see not only mass migration but also violent solutions to refugee problems, not only tensions over water or mining rights but also resource wars, not only religious conflicts but also wars of belief.

—Harald Welzer[3]

Visitors to the Four Corners region of the American Southwest often stop and visit the ruins at Mesa Verde. Located where the states of Arizona, New Mexico, Colorado, and Utah meet, the Four Corners is a semiarid land rich in beauty and Native American history. It is a region of high flat mesas bisected with long, deep slickrock canyons, sandy washes, and scattered clumps of sagebrush, juniper, and small groves of cottonwood trees clustered around usually dry riverbeds. This rugged landscape makes navigating the area tortuously hard as you can never travel in a straight line and are constantly climbing and descending and changing direction in order to avoid cliffs, washes, and the like. Varying in elevation between 6,000 and 8,500 feet, the Four Corners are a transition zone between the San Juan and La Plata Mountains to the east and north and the lower desert terrain to the west and south.

Richard Wetherill and his brother-in-law, Charlie Mason, were two ranchers hunting for stray cattle in this rugged country on December 18, 1888, when they became the first non-Natives to discover the ancient cliff dwellings of Mesa Verde. They were riding along on the top of a high mesa that covers around 520 square miles of uplifted tableland in southwestern Colorado searching for their missing steers. Traversing this landscape in December was not for the faint of heart, and at one point driving snow forced Wetherill and Mason to dismount in order to avoid riding over a cliff edge in the snowstorm. As they slowly made their way forward they came upon one of the deep canyons that cleaves the southern slopes of the mesa. Incredulously, as they peered through the snow at the chasm beneath them, they glimpsed a stone city halfway up the facing canyon wall nestled in a protective alcove. Sheltered by a massive sandstone overhang, this cliff palace, as they subsequently called it, boasted buildings three stories high and was in excellent condition. Giving up the search for the missing steers, the two men laboriously made their way down and across to the cliff below the alcove before they were finally able to climb up to the buildings. They spent hours exploring the ruins. This was the start of a new passion for Richard Wetherill, who spent much of the remainder of his life exploring the region and looking to discover the ruins and artifacts left behind by those who had once inhabited this region.

Over the years, I have taken a number of family trips to the ruins and each time have come away with new respect for the people who had once made their homes there. A national park today, the modern visitor drives up out of the flatlands from the nearby town of Cortez onto a mountain road that dramatically winds its way up the sides of the massif and then meanders along the top of the mesa giving access to a visitor's lodge, bookstore, and a number of separate cliff dwellings bearing names such as Cliff Palace, Balcony House, and Long House. If you purchase tickets for a guided tour at the visitor center, you can descend to some of these ruins accompanied by a ranger and the usual assortment of tourists. It's not easy to get down to some of the ruins, even with the engineering done by the National Park Service to make these ancient communities more accessible to the casual visitor. You have to negotiate ladders, narrow little pathways, chutes, and even a tunnel or two. Leaving one ruin, you actually climb up a cliff face and must partially rely on hand and foot holds to ascend the sheer wall, albeit with a railing close at hand for safety. Perched high up on these inaccessible cliff faces and under a sheltering overhang, you realize that it took a great deal of effort to construct these communities and a lot of inconvenience to live in them. There's no easy way in and out and the only food and water available is what you are able to carry in. Perhaps the inaccessibility is the whole point. Although much is still not known about the people who lived there and why they built these high and remote communities, researchers have been able to piece together much of the history of the region and the people who lived there. The story that they've uncovered tells us a great deal about the connections between climate change and the origins of conflict. Even though this story took place long ago, it still has much to teach us about the ways in which climate change can stress communities, weaken social order, and increase the risk of widespread conflict and violence. The nature and type of violence that may develop out of climate change in the present day are the focus of this chapter, yet the story of the Mesa Verdeans can serve as an object lesson from history as to the connection, or as the Bard famously wrote in *The Tempest*, "What's past is prologue."

The Native people who built the cliff dwellings at Mesa Verde and many other well-known ruins in the Four Corners region—such as Chaco Canyon—are popularly referred to as the Anasazi. It is a Navajo term often interpreted to mean "ancient ones" but is better translated as "ancestors of our enemies."[4] No one knows exactly when the first peoples arrived in the Four Corners region, but the Anasazi culture, which developed from the Paleo-Indians who preceded them, was largely made possible by two new developments. The first was the onset of a period of more regular precipitation from 2000 BCE to about 500 BCE, while the second involved the introduction of maize and then beans.[5] Previous to these changes, small family bands roamed throughout the region for countless generations eking a precarious life from the limited offerings of the region. Marked by dry washes, scrub brush, and sandstone mesas, this semiarid landscape is a hard place to survive given the often harsh weather that can veer between extremes of heat and cold. More importantly, it also suffers from a chronic lack of water. Limited supplies of game and sparse vegetation meant that the Natives of the region had to work hard and move often to keep themselves alive. Sporadic rainfall and limited food also meant that life was often precarious and difficult. But survive they did. For generations they lived and traveled in small family bands well adapted to such an environment. Smaller groups could more easily pull up stakes and move in the perpetual quest for game and forage and fewer mouths meant that it was easier to find enough food to feed everyone in a landscape in which adequate sustenance was often hard to come by. Dispersed in small wandering groups, the human inhabitants of the Four Corners region were scattered and few in number. The land simply could not sustain larger population groups, but a changing climate resulting in increased precipitation over the region, coinciding as it did with the introduction of agriculture, changed life in the Four Corners region in dramatic fashion.

As the region entered a wetter period, populations flourished because increased precipitation meant more game and edible plants. This process only accelerated after the Paleo-Indians developed a greater reliance on agriculture—largely fostered by the introduction of maize, followed by beans—which ended the seminomadic lifestyles that they and their ancestors had practiced for generations. Because of this larger dependence on farming, the population of the Paleo-Indians exploded as food supplies became more plentiful and reliable, especially as farming skills and techniques were refined and developed that allowed them to maximize crop yields. All of these changes facilitated the evolution of a more complex and sophisticated culture that we now refer to as the Anasazi and soon communities proliferated through the region. Corn wasn't always easy to grow as it took time and effort to cultivate this crop, but it was worth it since in a region notorious for fickle weather, stored corn provided a good source of food during winter and other lean times. Even basic forms of agriculture can support and nourish around fifty times more people than foraging can.[6] Previous to the introduction of agriculture, starvation and famine were not unknown during especially hard winter seasons, but with greater food security, communities swelled in size and number throughout the Four Corners region.

This turn to agriculture, however, for all of its benefits, also came at a cost. Even while providing more consistent and nutritious food, it also meant a greater vulnerability to climatic variations and uncertainty. In writing about the transition from foraging and hunting to agriculture, noted scholar Karen Armstrong points out:

> In our industrialized societies, we often look back to the agrarian age with nostalgia, imagining the people lived more wholesomely then, close to the land and in harmony with nature. Initially, however, agriculture was experienced as traumatic. These early settlements were vulnerable to wild swings in productivity that could wipe out the entire population, and their mythology describes the first farmers fighting a desperate battle against sterility, drought, and famine.[7]

This provides an apt description of the situation in the Four Corners region of the Southwest. Even during the best of times the rains were still relatively sparse and haphazard. During the monsoon season, one community might get plenty of rain while just a few miles away another would get nothing. Every year the villagers must have waited for the summer rain season with a mix of hope and dread: hope because the rains promised a good crop and enough food to survive, and dread because if the rains didn't come, the community faced empty bellies and a real risk of starvation. The late summer monsoon was especially critical since it coincided with the vital tasseling stage of the corn. People, however, can be adaptable and, over time, the patchy nature of the summer rains helped create a vast system of trade throughout the region. A community that had not received any rain might exchange things like baskets and bowls with other communities that had received rainfall and which might have some extra corn to barter. In this way, the environment helped shape an extensive economic system that tied the communities of the Four Corners region together in a web of trade and interdependence.

Over many years, the variable weather patterns in the Four Corners region resulted in a societal transformation that came to be character-ized by numerous small farming villages centered around larger com-munities that served as religious, trade, and cultural centers. Some of these dwellings, such as Pueblo Bonito in Chaco Canyon, boasted more than six hundred rooms. In fact, Chaco Canyon appears to be where the great houses originated and where the greatest concentration of them was to be found. Eventually this pattern of settlement spread through-out the Four Corners region with larger hub communities serving as storage centers where surplus corn could be stockpiled for those small-er and needier outlying settlements that might have experienced a bad year. These hub communities also appear to have developed "sophisti-cated astronomical devices,"[8] in order to help determine the best times for planting and harvesting crops. Often referred to as the Chaco era, this period was characterized by the development of an increasingly stratified society with those in the great houses—the hub commu-nities—having more wealth, goods, and power, while those in the small farming communities led lives marked by hard work and few extra resources and goods. Even back then, wealth had it privileges. Anthro-pological research indicates that the mortality rates for those children born and raised in the great houses was much lower than those born in the smaller farming communities. But even so, the people as a whole prospered and the Chacoan phenomenon spread as far as Utah, Colora-do, New Mexico, Mexico, and Arizona.[9] Pilgrims flocked to the great houses in Chaco Canyon, the spiritual and cultural heart of this civiliza-tion, and marveled at the size and wealth of the great houses, participat-ed in ceremonies, and helped build new great houses.

More dependable food supplies resulted in more people surviving in any given year and as the population grew, pressure increased to spread out and cultivate more land in order to feed all of the extra mouths. Good land became scarce and, consequently, ever more marginal land was settled and farmed. Over time, local resources were exhausted and people had to travel ever farther distances to hunt game and gather wood for heating, cooking, and building. So scarce did timber become that trees were felled and brought back from as far away as the Chuska Mountains located around sixty miles from Chaco Canyon and the Zuni Mountains around fifty miles away.[10] Needing to transport wood from such distances could have been a powerful warning, but it is unlikely that it was heeded. After all, the Chacoans were a mighty civilization. But what they didn't realize was that their society was already at its limit and the success their people had enjoyed carried within it the seeds of their own destruction.

For a time, this system that had so slowly and painstakingly been created to meet the needs of life in the Four Corners region functioned quite well and allowed the Chacoan Anasazi culture to flourish, but this all changed with the return of drier conditions around 1090 CE. There had been other short-term arid spells that had resulted in some dislocation, but this new drought was to be of much longer duration with ultimately catastrophic effects. During these earlier periods of low rainfall, it appears as if the social elites responded with rituals calling upon the gods for more rain and had also embarked on building projects to employ those who had lost their farms and livelihoods. Many new religious kivas (specialized rooms, often sunken, used for religious and political purposes), roads, and great houses were constructed during these episodes. These were public works projects designed to maintain social stability by using the labor of the unemployed to build for the public good and to keep a lid on possible social unrest. But these solutions, while somewhat successful in the short term, did not address some of the underlying weaknesses of the society: too many people in a landscape of scarce resources. When a long-term climatic change settled on the region and the rains failed to come, year upon year, the old tried and true methods for dealing with shortfalls were attempted, but always with diminishing success. Ultimately when the grain silos ran out and widespread famine became common, these strategies proved futile. Eventually, the inevitable happened, and the social structure collapsed piece by piece in a cascading avalanche of disorder, dislocation, and conflict.

In modern terms, it might be framed as an example of state failure leading to civil war, anarchy, and forced population displacement: all themes to which we will return in this and ensuing chapters. The priestly class lost their authority and legitimacy in the face of their apparent inability to bring more rain. Farmers left their farms in ever greater numbers and the great houses shrank in size and in many cases were entirely abandoned. During the 1100s, large numbers of Chacoan Anasazi migrated to the surrounding uplands to the north where they quickly came into conflict with those already living there, while those who remained in the traditional Chacoan homelands struggled to survive and protect what little resources they were able to maintain. It was a period of widespread societal collapse and upheaval that frequently devolved into violence.

Archeological evidence from this period reveals a great deal of newly built fortifications, with many small farms and homes bearing clear signs of having been pillaged and burned out. Human remains uncovered from this period often bear the marks of extreme violence including dismemberment, decapitation, crushed skulls, and various other evidence of brutal death. Even evidence of cannibalism has been found. Genetic testing of fossilized feces known as coprolites has revealed evidence of proteins that could only have been derived from other human beings.[11] Whether the cannibalism was a response to starvation or if it also carried spiritual or religious overtones is unclear, but it does seem to be symptomatic of a wider societal decline into anarchy and extreme violence. As anthropologist David Stuart succinctly notes in regards to the remains uncovered by archeologists:

> They were the desperate conclusions of hand-to-hand fights among people struggling either to acquire land and foraging territory before they starved (in the case of the Chacoans) or to hold onto it (in the case of the indigenous residents) merely to support their own families. An estimated 60 percent of adults and 38 percent of children died violently in the Gallina highlands after the collapse of Chacoan society.[12]

For at least three generations, this region was characterized by widespread turmoil, conflict, and violence. It was during this time of chaos and warfare that the cliff palaces at Mesa Verde were constructed. Sheltered and protected under massive sandstone overhangs, these communities provided safe havens for refugees with the added benefit of also providing security for any surplus grain and other food supplies that needed to be stored until consumed. They were also relatively warm in winter since most were built in the alcoves of south facing cliffs and received a good amount of winter sun. Most of the cliff dwellings had small farm holdings above them on the mesa top where many must have lived during the planting and growing seasons, but during winter or when in danger of being raided, the cliff dwellings provided a nearby sanctuary to wait out the bad times in relative safety. Interestingly, these cliff castles were only inhabited for about seventy years between the 1190s and the 1260s. While we don't exactly know why these cliff dwellings were abandoned, many believe that once again it had to do with climatic changes. During the 1260s and 1270s rainfall patterns again decreased in terms of frequency. This coincided with a period of prolonged below-average temperatures that also served to diminish the growing season for crops.[13] Under pressure from these twin blows, the inhabitants found that dryland farming practiced in the uplands around Mesa Verde could no longer sustain them and so, once again, climate change contributed to a pattern of increased pressure, social change, and ultimately conflict.

At first, it appears as if the inhabitants of the region around Mesa Verde shifted from growing maize to hunting and gathering as they struggled to diversify their food supply, but as life got harder people started to leave the region. Ultimately, as adaptation strategies failed, the area descended into more and more violence, which further hastened the regional exodus.[14] A great deal of archeological evidence from the region around Mesa Verde indicates that during this era there was a proliferation of defensive architecture and burned out buildings and communities. Human remains from this period reveal evidence of violent attack and death, scalping, and a lack of postmortem care. In other words, skeletal remains uncovered in the ruins of burned settlements often had evidence of lethal traumatic violence and the fact that the bodies of the victims were left in situ suggests that friends and families were also either killed or had fled in haste without having had the time to engage in the normal funeral rituals for fallen loved ones. While much of this violence was likely between neighboring settlements, it's also possible that locals were increasingly competing with newcomers to the area who may have themselves been displaced because of climatic changes, such as the recently arrived Navajo, Ute, and Apache.[15]

What happened to these migrants? Most believe that they dispersed throughout the Four Corners region, especially to the south and the east, and eventually became the Puebloan peoples that still live in the Four Corners region to this very day. In fact, many modern Puebloan tribes, such as the Hopi, Zuni, Jemez, Taos, Cochiti, Acoma, and Jemez, trace their ancestry back to the Anasazi in their oral traditions. In summary, the history of the American Southwest reveals how, in the past, climate-induced stressors accumulated over time and resulted in dislocation, societal collapse, and extreme violence among the Native peoples living in the region as their political, religious, and social structures proved unable to adapt and cope with the pressures brought about by population growth, limited natural resources, lower average temperatures, and a long-term decrease in available water supplies. Tried and true responses that had worked previously failed to meet the longer-term stresses and challenges. Ultimately, the civilization collapsed into violence and anarchy. Climate change created the conditions necessary for the creation of the Chacoan civilization, yet it also doomed that same society to a violent end when conditions reverted again and the societal structures and practices proved unable to meet the needs of an altered environmental landscape. Long-term cycles of drought interspersed with periods of increased precipitation are the norm in the American Southwest, and such an environment is unforgiving when populations outstrip resources. There are important and relevant lessons here that are hard to ignore.

The causes of conflict and organized violence are the same as they ever were. Novelist William Faulkner once wrote, "The past is never dead. It's not even past."[16] The example of the Four Corners region attests to the truth of those words. Even though the story of the Anasazi, Chacoan, and Mesa Verde peoples occurred in the distant past, the fundamental problem remains the same and is still instructive, even in the modern age. Climate change creates stressors that accumulate over time and undermine the ability of a population and its leaders to cope and adapt. The societal disorder, collapse, and violence that result from the inability of a society to effectively meet the consequences and challenges of climate change remain very real risks. Increasingly, we are waking up to that reality.

On July 23, 2015, for example, the Pentagon issued a report asserting that "climate change is an urgent and growing threat to our national security, contributing to increased natural disasters, refugee flows, and conflicts over basic resources such as food and water. These impacts are already occurring, and the scope, scale, and intensity of these impacts are projected to increase over time."[17]

The report further warned that "case studies have demonstrated measurable impacts on areas vulnerable to the impacts of climate change and in specific cases significant interaction between conflict dynamics and sensitivity to climate changes."[18] The Pentagon, though, is hardly the first to highlight this connection. As far back as 1991, Ian Rowlands wrote that "no country will be immune from the security challenges posed by global environmental change,"[19] while in 2004, a United Nations report noted, "Poverty, infectious disease, environmental degradation and war feed one another in a deadly cycle."[20] As these reports emphasized, climate change poses some clear threats to human communities by creating conditions conducive to the development of various forms of conflict. Examining and discussing some of those risks and conditions are the fundamental focus of this book. Just as the Chacoan people struggled to cope, adapt, and survive, so too will modern societies struggle. While some will succeed, others may not. For those communities and nations unable to negotiate the challenges of climate change, the price of failure may be just as high as it was for the Chacoan civilization long ago.

Violence, in its many forms, is likely to be the cost of failing to manage the challenges posed by an altered environmental landscape with all of the attendant political, social, and economic pressures. The point of such a recognition is not to suggest that violent conflict is inevitable, but rather to assess the ways in which the projected consequences of climate change will, in some locations, foster conditions in which violent struggle and conflict become much more likely. History is filled with examples of just such violent outcomes in the wake of profound climatological, social, political, and religious change. The violence itself has taken many different forms and will do so again. Sometimes the violence will assume ethnic, racial, religious, or nationalistic overtones, but at heart, each will represent a form of collective aggression enabled, encouraged, and facilitated by individual and communal responses to the stresses and challenges posed by climate change. In the next chapter, I will discuss a number of specific pathways and processes that link such violence to climate change, but for now it may be useful to first distinguish between different types of collective violence in order to better understand the nature and dynamics of these destructive reactions and responses.

Violence arising due to climate change will take a variety of forms and occur for a number of different reasons. Violence tends to be highly contingent and situational; that is, it is largely shaped and influenced by the specific decisions of political and military leaders, and by local traditions, contexts, and circumstances. Although the violence itself may vary, we should not make the mistake of seeing the different types of violence as completely separate and distinct from each other. Differing in terms of the degree of spontaneity, planning, organization, and duration, they nevertheless represent related forms of group violence against members of other groups. These forms of violence will also share a common genesis in stemming either directly or indirectly from climate change impacts. One way to look at such varied, yet connected, forms of group violence is to imagine them on a continuum of violent conflict. At one end of the spectrum are examples of somewhat sponta-neous ethnic and communal violence such as riots and pogroms, while at the other end lies the organized, planned, and systematic violence of both war and genocide. Along the way are countless permutations and variations that include uprisings, peasant rebellions, warlordism, terror-ism, revolts, coups d'état, and any number of other manifestations of group violence. At different times and places one form may morph into or give rise to another—war into genocide, for example—and any num-ber of these types may occur concurrently with each other.

At the spontaneous end of the spectrum, the violence of riots and pogroms will be largely driven by emotional reactions to fear, anxiety, and frustration at the unsettled and difficult nature of the times and because of perceived threats. Such unplanned and unstructured vio-lence will be especially likely where preexisting and long-standing prej-udices against specific minority groups already exist. Riots and pogroms can be understood as being expressive types of communal violence. Those who study violence often distinguish between those types that are largely expressive in nature, as opposed to those that are primarily instrumental in orientation.[21] Expressive violence concerns those in which the motivations are indicative or "expressive" of some emotional state, such as fear, rage, anger, or jealousy. Riots, for example, are largely expressive in that they tend to develop fairly spontaneously and reflect emotionally laden and violent reactions to perceived threats, injustices, or difficult circumstances. Instrumental violence, on the oth-er hand, refers to those forms of violence in which the destructive action is intended to achieve some aim. The violence in these cases is a tool or "instrument" in accomplishing some other goal. War and geno-cide, at the more organized and planned end of the spectrum, can be seen as being more instrumentalist in orientation. To further assess the nature of these different, yet related forms of group aggression, we can begin with riots and pogroms, which represent the most spontaneous and least organized end of the continuum, before moving on to a dis-cussion of war, and then finally genocide.

Riots can be broadly defined as occurring when groups of people react to a specific trigger or grievance and engage in violent and de-structive actions. The pattern is usually the same. Most riots start with a trigger. Something happens, in other words, that attracts a group's at-tention and interest, after which the focus of the crowd or mob begins to converge around a common element as emotions strengthen. United around an issue, individuals within the group start engaging in behavior that is quickly mirrored by others. It might be as simple as someone throwing a rock, smashing a car window, or attacking someone. Others in the crowd, caught up in the moment, also unleash their aggression and begin acting violently and a riot is born.[22] Buoyed by the sense of power engendered by a crowd and freed from individual normative constraints by a diffusion of responsibility and sense of deindividuation that being an anonymous member of a large group provides, individuals in these contexts are relatively easily moved by the violent passions and actions of others.[23] Riots, simply put, are a largely spontaneous expression of mob violence, although they can sometimes be orchestrated and set in motion by those trying to achieve some political or social goal. In such a case, that riot could be seen as being both expressive and instrumental at the same time. Riots can be directed against any number of targets and occur for many different reasons, some as trivial as losses on the soccer field or basketball court, but more commonly around consequential issues such as race, religion, and ethnicity.

In contrast, pogroms, while very similar in some ways to riots, tend to be larger scale and longer lasting. They also usually involve members of a majority population targeting members of a minority group.[24] The term itself was first used to describe attacks against Jews in the Russian empire during the latter years of the nineteenth century and the early years of the twentieth, but since then it has also been commonly applied to describe attacks against many groups, not just Jews. Both riots and pogroms may sometimes evolve and change as they develop or as circumstances and reactions evolve in response to the initial violence. While riots may arise somewhat impulsively and spontaneously because of fear and hysteria, especially if they build upon a previous history of intolerance and prejudice against a group, it is also equally true that such destructive behavior may constitute a manufactured response set in motion by those who stand to gain from the violence. Political, social, and/or religious leaders sometimes mobilize populations in order to scapegoat vulnerable groups, act on preexisting prejudices, divert attention away from failed policies, or enhance the solidarity and internal cohesiveness of a group by targeting outsiders. Not uncommonly, riots and pogroms may overlap and evolve, such as when ethnic violence breaks out somewhat unexpectedly and is then capitalized upon by various leaders who see in the outbreak of violence opportunities to advance their own agendas. The ethnic and religious violence that has sometimes broken out in India illustrates this kind of scenario well.

In 1984 anti-Sikh riots broke out after the assassination of the Indian prime minister Indira Gandhi by two of her Sikh bodyguards.[25] Gandhi had previously ordered the army to retake the Golden Temple, a Sikh holy site that had been occupied by Sikh separatists. Sikhs are an ethnic religious group found throughout India but predominate in the northern Indian state of Punjab. When the military complied with Gandhi's directive they did so in a heavy-handed way that resulted in a great loss of life,[26] and Gandhi's assassination was directly motivated by her order to send in the military. After her death, riots broke out against Sikhs throughout India that were marked by wholesale assault and murder on the largely defenseless Sikh minority. The violence also had a gendered component as well, since Sikh women, targeted as both women and Sikhs, were specifically singled out and subjected to systematic and widespread gang rapes. Such intersectionality is not uncommon in such forms of collective violence. These largely spontaneous and localized outbursts against Sikh minority communities were soon aided and abetted by Hindu nationalist political and police officials who exploited the situation and provided a quasi-official sanction for the violence.[27] Thousands were killed and many more injured before the arson, killing, rapes, and assaults were finally brought to an end. Unfortunately, such ethnic and religious violence has often marked India's history,[28] although more typically it has involved Hindu action against the Muslim minority such as occurred in 2002 when a train filled with Hindu pilgrims caught fire accidently. Because of a long history of prejudice against Muslims and previous bouts of violence against them, the Muslim community provided a ready-made scapegoat and in the aftermath of the train fire widespread anti-Muslim riots broke out involving much loss of life and destruction of property.[29]

Ethnic and religious riots are not unique to India and both spontaneous and coordinated examples of communal violence have appeared in such far-flung places as the Caucasus,[30] Burundi,[31] and the Philippines[32] among many other locations. Old prejudices often provide fertile ground for violence in response to contemporary events and grievances. Increasingly, such destructive processes may, at some level, also revolve around climate change. As destructive and horrific as these riots, massacres, and pogroms have been and as bad as they might get, they pale in comparison to the scale and systematic nature of warfare, one of the oldest examples of group conflict.

No less a figure than Thomas Hobbes, the famed English political philosopher and author of *Leviathan*, believed that war is the natural state of humanity and that most of human existence involved people living in "continual fear, and danger of violent death." He then went on to famously suggest that because of this situation, life was "nasty, brutish, and short."[33] War has been such a ubiquitous companion to human history that some have wondered if war isn't in fact part of human nature.[34] Certainly, if we look at our closest genetic relatives, the chimpanzee, we find that contrary to the image of a peaceful primate, they are often violent and engage in murder, rape, assault, and cooperative acts of violence.[35] In some ways it is possible to perceive warfare as a legacy of our evolutionary heritage. Human beings evolved in a world in which violence often proved advantageous for survival and warfare may be a natural extension of our propensity for violence that developed as human communities grew and came to understand that collective and organized violence offered an effective means of acquisition and defense.

Best estimates suggest that the advent of war dates back to the Neolithic era approximately ten thousand years ago,[36] when the development of agriculture and the resulting growth in the size and complexity of settled human communities forever changed the nature of conflict.[37] Interestingly enough, recent research indicates that the introduction of agriculture actually altered the DNA of human beings, impacting the height, digestive and immune systems, and skin color of humanity.[38] The introduction of agriculture was also to have important ramifications for organized violence. While there is ample archeological evidence of massacres and other forms of group violence previous to agricultural modes of subsistence, the level of social organization needed to create and maintain armies simply did not exist. The fighting that took place was only loosely coordinated and typically involved ad hoc formations of young men from a given clan, tribe, or other kinship-based group. But with the advent of farming and the consequent production of larger and more regular supplies of food, populations settled around arable land and grew in size and density. Over time, the ability to grow crops was refined and allowed for the development of food surpluses that needed to be stored, and as a consequence, communities grew larger and more populous. As they grew, social, political, economic, and religious structures were developed in order to meet the increasingly diverse needs of a growing community and to provide mechanisms to allocate services and resources, maintain social order and harmony, and provide meaning and identity. This flowering of social complexity and organization was the start of civilization. It was also the beginning of organized warfare since the surplus food and greater material wealth fostered by agriculture sometimes proved to be an attractive target for jealous or needier neighbors.

The evidence suggests that the first wars often centered around attempts to steal supplies and foodstuffs from more affluent communities.[39] The first wars, in other words, were resource wars. It is quite probable that they involved bands of nomads attacking settled communities either during or after the harvest since those settings offered the prospect of easy pickings at a known location. Naturally, these cities soon developed defensive fortifications such as walls and ramparts in order to protect themselves and their resources, but soon enough these same cities learned that they could attack each other and profit from wars designed to gain resources, slaves, and other goods.[40] In short, the evidence suggests that some of the earliest known examples of war were struggles over resources: some driven by need, some by greed.[41] It also tells us that even at the dawn of history, war clearly involved an element of rationality and calculation.

War is a means to an end, a more or less reasoned strategy intended to accomplish some purpose. All wars have, with varying degrees of success, served a variety of goals and objectives that reveal some sort of underlying rationality. Certainly, wars have been fought over miscalculations and mistakes, as well as for belief systems, ideologies, and religion; and just as certainly irrational ideas, such as racism, nationalism, and xenophobia, have often shaped the outcome of these decision-making processes. These factors have often coalesced in a toxic mix of calculation, greed, prejudice, arrogance, pride, and hatred. Lest we believe, however, that such thinking is aberrant or the result of mental disorder, we need to remember that decisions leading to war are often supported and even encouraged by large segments of the population. The outbreak of World War I in August 1914, for example, was greeted with widespread acclaim and celebration throughout Europe as many thousands saw in warfare an opportunity for national glory, redemption, and individual heroism.[42]

Wars start because decisions are made, policies initiated, and troops and populations mobilized when it is seen as a logical, even a desired, means to an end. In other words, wars don't just happen. They are instead the result of choices made by political, military, and/or religious elites intent on achieving some desired goal or goals and who decide that the costs and risks of such a conflict are outweighed by the possible benefits. These more pragmatic deliberations are often disguised, distorted, supported, justified, and buttressed by a wide variety of ideological, religious, and nationalistic motivations and justifications, but they are present nonetheless. Whether we are talking about interstate war, civil war, low-intensity war, wars of colonization or decolonization, or any other form of warfare, this element of conscious decision making is invariably a part of the complex mix of considerations that brings such conflict into being. When we consider this in light of the projected consequences of climate change on human communities, it paints a dark picture of disturbing possibilities. The study of history teaches us that wars and other forms of collective conflict have often revolved around resources, territory, power, nationalism, intolerance, and fear.[43] It is therefore critical to ask whether or not these historic patterns will reemerge as states and population groups confront resource loss, state failure, deprivation, population dislocation, and other climate-induced changes and emergencies in the coming years. Clearly, much will depend on the choices and calculations of social elites and population groups.

While riots are largely spontaneous responses to a specific trigger, war and genocide are much more deliberate strategies intended to solve a problem, achieve some goal, or respond to a perceived threat. While international wars have been in decline since the end of World War II, it is also true that we have seen a proliferation of smaller conflicts, especially civil wars in which neighbor is pitted against neighbor, and community against community within the same nation.[44] In fact, civil wars, such as those going on in Syria, South Sudan, Somalia, the Central African Republic, and Iraq, now represent the most common type of large-scale group violence.[45] These types of internecine conflicts are often more typical in poor nations with weak and/or corrupt governments and with an economy that largely relies on export of some primary commodity.[46] Such political and economic conditions open the door to corruption, exploitation, and the ability to finance rebellions and insurgencies. Civil wars are also more likely in multiethnic states in which one group is largely dominant and are often the result of some combination of what has been termed "greed and grievance,"[47] in which resources and intergroup hostility serve to make violent conflict and competition more likely. Will climate change create conditions making violent struggle more likely? Will we see oppressed minority populations employ revolutionary and violent solutions to their problems or dominant ethnic groups use repression and state terror to maintain their hold on power? Keep in mind that those societies prone to civil war tend to be those nations most vulnerable to climate change with the fewest means for dealing with projected impacts. Such wars will also be fought for a variety of reasons that may not appear to have anything to do with climate change, yet underneath the trumpeted justifications and religious, ethnic, or racial dimensions, many will be waged as a consequence of the challenges and societal stresses created by a warming world.

In addition to the heightened risk for riots, pogroms, and different types of warfare, especially civil wars, it is also my contention that the consequences of climate change will pose a particular risk for the development of what can best be termed the genocidal impulse: the attempt to exterminate an entire population group. Genocide is a particular form of collective destructive action that largely targets defenseless populations because of their religion, ethnicity, race, or nationality. This is why the political scientist and genocide scholar Martin Shaw, for example, refers to genocide as a form of degenerate war since, in many ways, it represents a war against civilians.[48] Predictions suggest heightened competition and conflict over increasingly scarce resources and opportunities and millions of people being forcibly displaced because of the various consequences of climate change. The combination of vulnerable populations during times of change and increased competition is a recipe for violent conflict and the development of the genocidal impulse.

Originally coined by Raphael Lemkin in his 1944 book, *Axis Rule in Occupied Europe*, the concept of genocide was intended to capture the scale and nature of the Nazi crimes in occupied Europe.[49] To Lemkin's way of thinking, other terms such as "war crimes," "barbarism," and "massacres" did not capture the full nature of the systematic, comprehensive, and all-encompassing nature of the Nazi assault on the Jews. Accordingly, he created the word from the Greek *genos*, which means race or tribe, and the Latin *cide*, which means to kill. Genocide, therefore, refers to the attempt to kill a population group. After the end of the war, Rafael Lemkin worked with the newly created United Nations to turn his concept into law and was instrumental in finally getting it defined as a crime under international law in December of 1948.[50]

The concept of genocide was crafted in such a way as to incorporate a broad assortment of processes and policies, some more self-evidently destructive than others. It is purposeful in orientation and encompasses a range of destructive tactics, not all of which are overtly violent and not all of which are direct forms of destruction. Genocide can range from short-term measures that involve direct physical killing to longer-term and less straightforward strategies intended to eliminate a population in the future by removing children or preventing new births from within that group.[51] In many ways, genocide represents the far end of the continuum of organized and collective violence discussed earlier in this chapter, although genocide does stand apart in terms of its sheer lethality. During the twentieth century, an era sometimes characterized as an age of total war,[52] genocides killed roughly four times as many people as all the wars of the twentieth century combined.[53] Despite such a distinction, genocide is closely connected with other forms of collective violence, especially warfare. In much the same way that war can be understood as a somewhat rational strategy intended to achieve some objective, so too does genocide represent a conscious and intentional means to some end. Genocides are at least partially perpetrated because political, social, and religious leaders decide that they represent a viable solution to a real or perceived problem, which has led the political scientist Benjamin Valentino to argue that to understand genocide, one must focus on the "strategic logic" of the mass killing.[54] Some genocides are perpetrated for belief systems, others for revenge, maintaining power, economic interests, or some combination of these motivations.[55] Those perpetrated for economic reasons, often known as developmental genocide, usually involve the forcible attempt to remove a targeted group perceived to be in the way of economic exploitation and/or colonization. Historically, we have most often seen this kind of genocide perpetrated against native peoples whose lands and lifestyles have been seen as impeding the ability of a state to settle a region, cultivate land, and extract valuable resources such as minerals, coal, and oil.[56] The violence of genocide, while at some level the result of conscious choices, is also deeply embedded in deeper and widespread societal forces and processes that can converge in various ways to produce this crime.[57] Various institutions and organizations, not just government leaders, can encourage and facilitate many types of violence and persecution for all sorts of complicated reasons and motivations. Genocide, in short, is very complex, occurs for multiple reasons, develops from multiple sources, and is often embedded within other forms of violence, such as war.

Genocide is often perpetrated during wars or in their immediate aftermath. This is not to suggest that genocides must take place during wartime, but in the modern era, they have usually taken place either during or immediately after one. All of the major genocides of the twentieth century, for example, were perpetrated within the context of wider wars that helped allow for the development and implementation of the genocidal impulse and the wholesale massacre of civilian populations.[58] It was World War I that set the stage for the Armenian genocide of 1915–1917,[59] while the Holocaust was enabled by the catastrophic violence of World War II, especially on the eastern front.[60] Similarly, the killing fields of the Cambodian genocide were facilitated by the spillover effects of the war in neighboring Vietnam and the consequent civil war in Cambodia itself.[61] The butchery of the Rwandan genocide also occurred in the midst of a conflict, in this case a civil war.[62] In each of these examples, war provided the milieu in which genocidal ideas and impulses could take root and be put into practice. While the precursors of genocide—things such as xenophobia, nationalism, and a sense of historic wrongs to name a few—typically predate the outbreak of war, the actual policies and practices of annihilation typically arise in societies waging war because of the nature of war itself. We can see this reality clearly if we examine the ways in which Nazi policy toward the Jews changed over time and resulted in the Holocaust.

Anti-Semitism was deeply embedded within the ideology of Adolf Hitler and the Nazis when they began their rise to political power in the 1920s. Anti-Semitic images and messages riddled Hitler's speeches, which, as the eminent British historian Richard Evans suggests, "reduced Germany's complex social, political and economic problems to a simple common denominator: the evil machinations of the Jews."[63] In many ways, such scapegoating allowed for the mobilization of a fearful, resentful, and deeply envious segment of German society.[64] The Nazis often used violence during the 1920s, against not only Jews but various political and social enemies as well, but this violence was not genocide. It was instead an expression of intolerance and hatred, a tool of intimidation and coercion, and was, in fact, part of the political culture prevalent throughout Germany in the post–World War I years. Even after achieving power in January of 1933 with the appointment of Hitler to the position of chancellor, genocide did not become policy for the new Nazi state. Certainly, the new regime persecuted and enacted laws to legally, politically, socially, and economically marginalize the Jews through policies of systematic discrimination, segregation, ghettoization, and emigration.[65] However, between 1933 and 1939, the Nazis never had a single consistent Jewish policy and even at one point considered deporting all German Jews to the island of Madagascar, far away off the southeastern coast of Africa.[66] It was only after the invasions of Poland in 1939, and especially the Soviet Union in 1941, that Nazi policy coalesced around extermination as the preferred outcome. Even after the invasion of Poland, however, the Nazis still pursued contradictory policies with local administrators in the occupied east and military commanders often working at cross-purposes as each sought to pursue their own particular agendas.[67] In fact, even in the months leading up to the invasion of the Soviet Union, the Nazis were still considering deportation as an option.[68] Even so, the Nazi occupation of Poland was extremely violent and brutal and served as a testing ground for persecutory racial policies and atrocities against Polish civilians, both gentile and Jew.[69] Much of this violence, however, was ad hoc and localized in conception and implementation. In occupied Poland, the German authorities increasingly struggled to adapt to the pressing issues posed by waging a war and dealing with the large Jewish communities under their control.[70] Poland alone was home to around three million Jews, while the eastern part of Russia, including the Ukraine, had around two and half million, with another quarter million in the Baltic states.[71] The violence in Poland, then, was only the prelude for what was to come in a more coherent fashion after the invasion of the Soviet Union. Nazi policy was to see a dramatic and deadly change with the next phase of the Nazi expansion eastward with a more systematic approach to violence against the Jews.

Operation Barbarossa was the code name for the invasion of the Soviet Union on June 22, 1941, and it set off the largest, costliest, most destructive, and deadliest war in human history.[72] Importantly, it also dramatically facilitated the radicalization of German attitude and policy toward the Jews. In January 1942, a number of leading Nazis and government functionaries met in a villa in Wannsee, an upscale suburb of Berlin, and began to coordinate the "final solution" to the Jewish problem, namely, the mass murder of Europe's Jews.[73] German authorities had finally decided on what came to be called the Holocaust; previous solutions had not worked because of the large numbers of Jews that were now living in German-occupied territory, and because of the nature of the war itself. It was a war that was defined by the Nazis as a national and racial struggle for survival between two competing ideologies and races: Nazism versus Bolshevism, and Aryan versus Slav and Jew. Furthermore, the invasion of the Soviet Union was built upon a belief that if Germany wanted to survive, *Lebensraum* or living space was needed.[74] It was also a war for resources.

Many leading Nazis saw the largely flat and fertile lands of eastern Europe as a viable location for German expansion and colonization and when planning for the war, government officials concluded, "If we take what we need out of the country, there can be no doubt that many millions of people will die of starvation."[75] Estimates were prepared that revealed twenty to thirty million people would starve to death because of Nazi policy, with one report concluding that "many tens of millions will become superfluous in these areas and will have to die or emigrate to Siberia."[76] In essence, the invasion of the Soviet Union was intended to provide land and resources necessary for the continued survival of the German nation and part of this process was to involve the starvation of millions of inhabitants of the conquered territories so that their food supplies could support the war effort and feed Germans. An added benefit would be that the newly acquired land would become available for colonization by ethnic Germans. From the very outset, therefore, the invasion of the Soviet Union was intended to be a genocidal war of survival, a war in which genocide was intrinsic to the goals of the conflict.[77] In an age of scarce resources and heightened competition for what remains, such cold-blooded calculations may once again play out with lethal effect.

We also cannot discount the role that the nature of war itself played in fostering a further radicalization of preexisting policies and tendencies among Nazi officials and officers.[78] Keep in mind that these tendencies are true for warfare more generally. Wars simply make genocide much more likely because they facilitate the development of extremist and genocidal ideologies and practices through a number of mechanisms inherent to warfare itself. They do so in a number of specific ways. First, war strengthens nationalistic feelings and activates a sense of national unity and purpose. The group or national identity is intensified as the collective sense of self is prioritized over the individual self. Psychiatrist Aaron Beck suggests that "in wartime the national image becomes the center of every citizen's worldview; as they rally 'round the flag,' they move from an egocentric to a group centered mode."[79] For members of the in-group it is a time of heightened belonging since the threat of warfare exaggerates communal tendencies and prioritizes ideological and behavioral conformity. The cost of such tendencies, however, is borne by those groups defined as enemies or outsiders, since they are vulnerable to being scapegoated and targeted. Any group identity is largely determined by the divide between "us" and "them." Inclusion, in other words, is largely delineated by exclusion. This tendency allows communities to come together in order to more successfully wage war, but it increases the risks for populations who "don't belong." The scholar Harald Walzer highlights this process when he asserts:

> It is the ambiguity or "viscosity" of group boundaries that can lead to extreme acts of violence in a crisis. The function of such acts is to establish once and for all who "we" are and who "they" are, who should be seen as a friend and who as an enemy. The violence itself draws the boundary line, since "we" and "they" are unmistakably clear after an attack or mass killing.[80]

Placing individuals and groups into a stigmatized class that does not belong is a way of weakening natural human tendencies to empathize and identify with other people.[81] It reduces human beings to things and allows them to more easily be defined as "the enemy" or as an "outsider." Targeting internal enemies for elimination is the essence of genocide and during times of war the dualistic thinking and identification of internal and external enemies heightens the risk that war-related violence may spill over or morph into genocidal violence. Engaged in fighting a war, a state can easily define an outsider or minority population group as the enemy and present the persecution and elimination of members of that group as a necessary, important, and legitimate service to the nation. The violence of genocide is simply transformed from a crime into a patriotic duty. This is a common sentiment among many genocidal perpetrators who define their participation as a patriotic service to their homeland.[82] Such a situation is more easily arrived at in locations where a history of previous conflicts and preexisting prejudices are already present.

In such settings, old antagonisms can be activated, exploited, and relied upon. Preexisting prejudice can provide a deep well or reservoir of hostility that can be exploited when needed. Derogatory belief systems facilitate the victimization of others since they serve to remove those targeted from the social obligations we instinctively give to others. Human beings are social creatures and have deep-seated ties, responsibilities, and commitments to other human beings. These connections and obligations are strongest for those closest to us—such as family, kin, and friends—and weaken the farther from us other people are perceived to be. The greater the social distance, in other words, the fewer obligations we feel toward others. Prejudice, therefore, serves to move those negatively depicted away from us in terms of perceived similarity and consequently lessens our sense of common humanity and kinship.

We also shouldn't discount the role that war plays in normalizing hatred of the "other." During times of war, normal modes of interaction and the rules of conventional behavior don't seem to operate anymore or are not as relevant. War, after all, represents the antithesis of peacetime principles. All of the things that are normally prohibited are suddenly allowed. During peacetime, violence is generally prohibited and law and social order are the norm. During wartime, however, these norms are reversed and certain kinds of violence become legitimized as a necessary evil. Intolerance of the "enemy" and violence become a civic virtue and an obligation and such attitudes are encouraged and supported by friends, neighbors, and social, political, and religious leaders. It's easy to see how these kinds of perceptions and mind-sets can extend into support for genocide. Think about how such processes may play out in a nation struggling to cope with the demands imposed by climate change and confronting violent conflict around resources and refugee populations. In such a scenario, it's easy to understand how violent conflict may radicalize and evolve into programs of expulsion and extermination.

This isn't to suggest that people lose their sense of ethics. People in wartime don't lose their moral sensibilities; they are simply replaced with a wartime morality—one, it should be noted, in which death and destruction are much more common and accepted. The ethicist Jonathan Glover puts it this way: "When mass murder is sufficiently reinterpreted, people can support it with an unimpaired sense of moral identity."[83] Examining genocide, it is clear that perpetrators often understand their participation as being justified and legitimize it with expedient morality. In an infamous three-hour-long speech given in Poznan, Poland, on October 4, 1943, Reichsführer Heinrich Himmler explicitly stated, "We have the moral right, we had the duty to our people to do it, to kill this people who would kill us."[84] This sort of sentiment is by no means unusual among those engaged in exterminating a population. Of this tendency, the genocide scholar Alex Hinton writes:

> Genocidal perpetrators seem to engage in similar cognitive shifts, as they overcome moral prohibitions against killing by dehumanizing their victims, displacing responsibility, and morally justifying what they have done (by diminishing the negative effects while placing greater emphasis on ideological beliefs that legitimate their actions).[85]

Both war and genocide share this quality in that the violence is typically portrayed in ways that serve to justify the destructive actions. It's also important to note that war tends to desensitize and numb people to death, destruction, and the suffering of others. Human beings can become inured in such a setting and this brutalization process also makes it easier for genocide to take place.

The psychologist and pioneering scholar of genocide, Ervin Staub, situates the relationship between war and genocide within the larger context of what he terms difficult life circumstances.[86] War may be a part of this equation, but not necessarily so. His argument suggests that economic difficulties, widespread political and criminal violence, and rapid changes to the social, industrial, cultural, and technological fabric of a society result in widespread dislocation, fear, frustration, anger, and a sense of threat. All, by the way, conditions likely to arise and develop in the wake of climate change impacts. These, in turn, lead to strong drives to protect the self, the community, tradition, culture, and way of life. The direction those drives take may vary, with differing risks for genocidal violence. As Staub writes:

> Constructive actions have beneficial, practical effects and also help a person cope with the psychological consequences of life conditions. Unfortunately, it is often difficult to find and to follow a practically beneficial course of action. When this is the case, it is easy for psychological processes to occur that lead people to turn against others.[87]

These tendencies, as we will discuss in the next chapter, can be exacerbated, exploited, and otherwise manipulated by social, political, and/or religious leaders who may share the fears, frustration, and anger of others and/or who may see in the situation a chance to further their own agendas. Yet even so, we need to remember that there is no single source of violence, whether it is individual acts or examples of collective violence. What we find instead is that violence is often the outcome of a host of contingent circumstances and possible responses. Each outbreak of violence tends to have a specific trigger or event that sparks the violence, but that aggressive response is also informed by a variety of complex and interrelated factors that include past history, preexisting prejudices and societal schisms, various ideological beliefs, and perceptions of victimization.

In an era of climate change as societies confront diminishing resources and population dislocation, and as conflict and warfare are seen as increasingly viable solutions or reasonable responses, social, religious, ethnic, and national cleavages can all too easily facilitate the development of the genocide impulse within the context of wider conflicts and war. These preexisting prejudices can stimulate the genocidal impulse since the essence of genocide centers upon the immediate or later removal of a population, usually a minority group, from a given geographic area and from within the social body of a national, racial, religious, or ethnic community. How exactly such responses may develop is the issue that we turn to in the next chapter.

3

LINKING CLIMATE CHANGE
AND CONFLICT

Preliminary research indicates that scarcities of critical environmental resources—especially of cropland, freshwater, and forests—contribute to violence in many parts of the world. These environmental scarcities usually do not cause wars among countries, but they can generate severe social stresses within countries, helping to stimulate subnational insurgencies, ethnic clashes, and urban unrest.

—Thomas Homer-Dixon[1]

Scapegoating, or picking on an innocent person, is a familiar example of displacement in the face of threat.

—John Medcof and John Roth[2]

People have very strong feelings, not only about their kin and close acquaintances, but about thousands, millions of remote and unknown people whom they have never met and may never meet. They may feel loyal and connected to those distant strangers, or conversely they may hate and despise them without ever having come across one of them.

—Abram De Swaan[3]

Pulling back the curtains covering the window in my hotel room, I was confronted with the skeletal remains of two tall office towers. Shell holes had been punched through the concrete sides of the buildings, while the front and back of these towering structures were largely gone, leaving many levels open to the elements. The buildings were simply gutted and had few windows remaining in place. Wires and cables dangled from the walls that remained and everywhere debris littered the floors. The year was 1998 and I was staying in a room at the famous Sarajevo Holiday Inn, a massive cube-like structure that had been built for the 1984 Winter Olympic games. This building had once held dignitaries, sports fans, and diplomats from around the world, but during the 1990s it became a powerful symbol of the violence wracking Bosnia after the end of the Cold War. Radovan Karadžić, the leader of the Bosnian Serbs, had often stayed here and held political meetings with his followers. In fact, it was from this hotel that snipers loyal to Karadžić fired on demonstrators in April 1992, an act many Sarajevans considered to be the start of the war. [4]

After the fighting broke out in earnest and the city itself was placed under siege, the Holiday Inn became the location from which many reporters established themselves and filed their reports. For three years, the city of Sarajevo was surrounded by hostile forces and the Holiday Inn was not immune to the sniper fire and artillery shells that fell on the city. In fact, the hotel was located in one of the most dangerous parts of the city. It sat on a stretch of road known as sniper alley linking the main part of the city with the airport; during the siege this road provided the only connection to the outside world and many Holiday Inn guests and staff became witnesses to the violence as it unfolded on the streets outside their rooms. After the violence ended in November 1995, with the Dayton Agreement, the Holiday Inn saw many aid workers replace the journalists as the city and country slowly began to rebuild.

In November 1998, just three years after the fighting ended, I was there in Sarajevo and staying in the same hotel that had seen such terrible events. At times, it felt surreal. Walking down the streets, I would sometimes see pockmarks on building walls and realize that these were bullet holes where people walking on the sidewalk had been shot at by some sniper during the siege. Other times, it was possible to see shrapnel scars on the sidewalks or roads. Later on, many of these would be filled with red resin and became known as Sarajevo roses because of the floral pattern thus created. That was in the future though, and during my visit the gouges remained ugly reminders of recent events. Crossing intersections, I would casually stroll across avenues where not that long ago pedestrians were risking their lives simply trying to cross the street. Looking up as I walked, I could sometimes see the mountains looming over the buildings. Sarajevo sits in a narrow valley in the Dinaric Alps and has mountains shouldering in from both the north and south. During the siege, crowds gathering for food and other supplies were visible to the besiegers on the hillsides above the city and were likely to bring down mortar or artillery fire. Snipers perched on these slopes had a ringside view of life within the besieged city and frequently shot at civilians going about their business on the city's streets and sidewalks. Ironically, some of the snipers were themselves from Sarajevo and after a day of shooting at residents on the city streets below them, would call their former friends in the besieged city to ask how they were doing and even reminisce about old times.[5] For almost four years, the citizens of this beautiful European city struggled and suffered with the realities of modern siege warfare and paid the cost in blood with almost 12,000 fatalities, and another 56,000 injured.

In the short time I spent in Sarajevo I came to question why this city had become the site of such appalling violence only a few short years earlier. How, I wondered, could it have been the scene of such extreme conflict during the closing years of the twentieth century? Sarajevo, after all, was a city with a long and storied history that lay at the crossroads of east and west. Founded in 1461 by the Ottomans, it was here that the Ottoman East met the European West and created a unique blend of culture, tradition, and religion that in the closing years of the twentieth century appeared to be emblematic of a tolerant, multicultural society. Yet when fighting broke out in 1992, the world was shocked at the speed with which Bosnia fell into the brutality and violence of what was chillingly referred to as "ethnic cleansing," a phrase resurrected from World War II when Croat fascists known as the Ustasha sought to rid their territory of ethnic Serbs.[6]

While the conflict in Bosnia was not a result of climate change, it nevertheless has much to teach us about the ways in which ethnic conflict, war, and genocidal violence can result from the wrong combination of circumstances, societal stressors, political leadership, fear, and intolerance. In an era of dramatic climate-induced impacts on nations, such variables may once again prove important for either enabling or hindering the development of violent reactions and solutions. As climate change increasingly impacts nations around the world, we will likely see the same dynamics at work in other places. The mechanisms and the social and political processes that facilitate war and the genocidal impulse are often the same, even though the underlying issues or particular triggers may themselves vary. This chapter is largely about tracing the specific pathways by which climate change impacts will translate into potentially violent and destructive outcomes and the example of Bosnia provides a textbook case for how social, political, religious, and ideological circumstances and processes can expedite and enable the development of violent conflict and the genocidal impulse. The case of Bosnia reveals many of the specific mechanisms by which societies descend into hostility, persecution, fighting, and exterminatory policies and is consequently worth discussing at length.

Bosnia, technically Bosnia and Herzegovina, was one province of the relatively short-lived state of Yugoslavia. Formed in the aftermath of World War I when diplomats drew up new maps after the collapse of the Austro-Hungarian empire, the country was created without much concern for geography or the regional distribution of different religious and ethnic population groups. The victorious powers simply carved up the carcasses of the Austro-Hungarian and Ottoman empires into nations whose newly created borders often cut across ethnic and religious boundaries.[7] Yugoslavia, or Land of the South Slavs, was crafted in such a way that it became home to a great variety of different ethnicities and religions including Croats, Slovenians, Serbs, Montenegrins, Albanians, Macedonians, and Bosnian Muslims. But it was not always a harmonious coexistence. During World War II, Yugoslavia was the scene of what one historian refers to as "wars piled one on top of another."[8] Germany and Italy invaded on April 6, 1941, and quickly defeated and dismembered the young state and for the next four years, guerrilla and antipartisan warfare left their mark on this defeated land. During those chaotic years, various ethnic nationalists exploited wartime conditions to settle old scores. Extremist Croat forces, for example, indiscriminately attacked and killed large numbers of Serbs, while members of rival partisan groups often fought one another as much as they fought the Axis troops stationed there. Estimates suggest that around one million Yugoslavs died during the war, the vast majority of whom were killed by fellow Yugoslavs, and such victimization was not easily or quickly forgotten.[9]

During the 1990s, Serb politicians often referred to the violence directed against Serbs during World War II as a genocide in order to drum up support for violence against other groups. In January 1992, for example, a number of Serbian Orthodox bishops warned Bosnian Serbs that they were living "under the threat that genocide will again be visited upon them,"[10] while in a television interview, General Ratko Mladić, who led the Bosnian Serb army and was responsible for the infamous massacres at Srebrenica, similarly asserted that the Bosnian Muslims were intent on the "complete annihilation of the Serbian people."[11] In this way, prominent Serbs were able to utilize the memory of past victimization and historic enemies as a powerful justification for their own aggression in the 1990s. Fear is a potent motivator for violence, but such references to historic victimization aren't just about drumming up and instilling fear. The psychiatrist Vamik Volkan developed the idea of "chosen trauma," to refer to a historic victimization that a group sometimes uses to define itself in the present.[12] Association with a past trauma becomes an integral part of the identity of that group and serves to reinforce and strengthen in-group cohesion. Sociologically speaking, we know that outside threats reinforce in-group solidarity and a "chosen trauma" serves as a potent reminder of a shared ordeal and grievance that can strengthen the bond uniting a people in addition to helping express their common identity.

For Serbs, their chosen trauma was a battle fought in June 1389 on a field known as Kosovo Polje or Field of Blackbirds. On one side was an Ottoman army and on the other was a force led by various Serb princes and while the battle itself was not decisive, it was subsequently described by many Serbs as a defeat that ended Serbian independence and led to an era of oppression. Over the years, this interpretation, though factually inaccurate, became enshrined in nationalist mythology, poetry, songs, and stories. While it might seem strange that a battle fought long ago could still exert such a powerful influence on the present, we need to understand that this sense of collective victimization was central to modern Serb identity; it was their chosen trauma. Referencing this battle, one scholar emphasized that "its real, lasting legacy lay in the myths and legends which came to be woven around it, enabling it to shape the nation's historical and national consciousness."[13] In the same vein, Louis Sell suggests that

> all nations shape their image of themselves, at least in part, on myth. For Serbia, the central myth is one of heroic struggle, often against hopeless odds, followed by betrayal and defeat, but also—eventually—rebirth and triumph. Like all national myths, the Serbian picture contains many exaggerations and downright falsehoods.[14]

Serb identity, in other words, has been inextricably linked with a sense of past loss and victimization that was to play a powerful role in the events to come.

The end of World War II saw the rise of one man who was to play a pivotal role in reuniting the Yugoslav people after the divisive and destructive fighting that had torn the nation apart. His name was Josip Broz, but he was better known by his partisan name Tito. He had been a successful guerrilla leader who fought the Nazis and subsequently took power after the defeat of Nazi Germany in 1945.[15] An extremely effective and often ruthless politician, Tito reestablished the Yugoslav Federation created in the aftermath of World War I, but which had been torn apart by the Nazis and partisan warfare.[16] Under a campaign for "Brotherhood and Unity," and building on his charisma and reputation as a war hero, President Tito worked hard to suppress ethnic identity in favor of a national Yugoslav one, and, in many ways, it appeared as if he had succeeded. To move past the ethnic nationalism that had so often marked the history of the region, Tito's regime tried to strike a delicate balance: respecting and protecting the rights of ethnic groups, while at the same time suppressing and criminalizing any political nationalism. The Yugoslav constitution explicitly prohibited "propagating or practicing national inequality and any incitement of national, racial, or religious hatred and intolerance."[17] Furthermore, Tito coupled these incentives with purges of Serbs, Croats, Muslims, Slovenians, Macedonians, and Albanians who advocated and pursued an ethnic-nationalist agenda.[18] For Tito, the only acceptable nationalism was a Yugoslav one. In forging this identity, he was aided by the fact that after breaking with the Soviet Union in 1948, he was able to position Yugoslavia as the only nonaligned communist state and adopted a more liberal and open form of communism than was found in the rest of the Soviet Bloc. Consequently, Yugoslavia received a fair amount of international aid and trade with the West, was open to Western European tourists, and enjoyed a level of peace, stability, and economic prosperity that was the envy of other communist nations.[19]

For all of his success in creating the Yugoslav state, however, Tito was not able to completely eradicate all of the deep-seated resentments and prejudices around ethnic identity that had played out so violently in the past. Of the changes wrought in Yugoslavia by Tito, the journalists Laura Silber and Allan Little write that "in retrospect the appearance of a stable and prosperous Yugoslavia may have been deceptive. Ethnic grievances had been suppressed, not dispelled."[20] This failure to eradicate ethnic resentments and prejudice was to have serious consequences in the 1990s. Essentially, Yugoslavia was a patchwork state with simultaneous elements of "co-existence and conflict, tolerance and prejudice, suspicion and friendship."[21] To those raised during the 1950s and 1960s, the common Yugoslav identity tended to be strongest, while to the older generation who had lived through the atrocities of World War II and to those raised during the 1970s and 1980s when ethnic identity again reemerged, there was little tolerance for those from other ethnic nationalities. Among the more urban and better educated, intermarriage and coexistence between groups flourished, while in many smaller, less educated, and less affluent communities, the old tensions persisted.[22]

Yugoslavia had been held together during the Cold War because of Tito's strong and charismatic leadership and because of a system of shared sovereignty between the six republics that made up the country. The intent was to prevent any one group from dominating the multinational, multiethnic state that was Yugoslavia, but real power rested almost solely with Tito and it all began to unravel after his death in May 1980.[23] Even though Yugoslavia was able to limp along for the next ten years, federal power slowly and inexorably began to deteriorate as the presidency rotated between the different republics that made up the country. This weakening of central authority allowed local communist parties in each of the republics to gain strength at the expense of national unity. It also reenergized the centrality of ethnicity to communal and political identity with a corresponding weakening of a common Yugoslav self. It certainly didn't help that Yugoslavia's economy during the 1980s was in bad shape with high rates of unemployment, declining living standards, and government austerity programs.[24] Although many didn't realize it yet, the nation was in crisis and as one scholar has noted, "Bosnia's history shows that although all three main groups have traditions of tolerance, extremism dominated in unstable periods."[25] Tragically, Yugoslavia was entering just such an era: one in which power-hungry politicians sought to capitalize on ethnic nationalism at the expense of the nation in order to try and retain their power base but succeeded only in dramatically heightening tensions between the various ethnic groups.[26] As the country entered the 1990s, these tensions gained ever more traction as each ethnically dominated republic sought to disentangle itself from the increasingly moribund Yugoslavia.

As we saw in the previous chapter, wars and genocides do not just happen. They are the result of decisions made, calculations and miscalculations acted upon, and a great deal of planning and organizing. Yugoslavia did not simply stumble into genocide and "ethnic cleansing." Instead we need to recognize that the driving force in the events leading up to the violence were self-serving and power-hungry politicians willing to exploit nationalist sentiments and the latent prejudices that had lain dormant in Yugoslavia since the end of World War II. One commentator put it very simply when he asserted that "Yugoslavia marched into hell because its leaders took it there."[27] As power increasingly devolved to the individual republics and away from the Yugoslav federal government in the years after Tito's death, a number of former communist politicians saw the writing on the wall and reinvented themselves as ethnic nationalists and began fostering policies based largely on ethnic nationalism. They exploited the fear, suspicion, and insecurity experienced by many citizens who saw all that was familiar and reassuring slipping away from them and instead offered a simple and powerful message of ethnic pride and national resurgence. Of particular importance during this time was Slobodan Milošević, a career politician with a rather undistinguished track record spent climbing the ranks of the Communist Party system in Tito's Yugoslavia and who was one of those who had reinvented himself as a nationalist. Franjo Tudjman of Croatia was another former communist who embraced nationalism as the key to power, but it was Milošević who bore the lion's share of responsibility for the genocide that was perpetrated in Bosnia.

In 1987 Milošević was sent to Kosovo to mediate the rising tensions between the majority Albanian and minority Serb populations, but instead of defusing the situation, however, Milošević appealed to Serb nationalist identity and promised that "no one should dare to beat you again."[28] The crowd of aggrieved Serbs loved it and with those words Milošević found the key to maintaining and even bolstering his power in the face of a waning communist system.[29] Things came to a head in the early 1990s as the Cold War ended and communism, as a political and national force, lost much of its vigor and relevancy. The first to go were Slovenia and Croatia, overwhelmingly ethnically homogenous republics. Economically prosperous, they had the closest ties to the West given their respective locations and were increasingly wary of being dominated by the powerful republic of Serbia—the largest and most populated of all the Yugoslav states—and its bellicose leader, Slobodan Milošević. Slovenia and Croatia declared independence in June 1991 and very quickly Milošević sent in the Yugoslav National Army. While the fighting was relatively short lived in Slovenia, about ten days in all, the conflict in Croatia was of much longer duration and involved a great deal of violence directed against civilian populations. Lasting over six months and marked by mass atrocities and war crimes, the conflict in Croatia resulted in more than 200,000 refugees, over 350,000 displaced persons, and around 20,000 killed.[30] In many ways, however, this was just a prelude to the violence that was still to come. Initially, it appears as if Milošević was mostly concerned with preventing the breakup of Yugoslavia, but as the inevitability of the dissolution became evident, he became more and more focused on carving out a greater Serbia from the rest of Yugoslavia. This was to be accomplished largely at the expense of Bosnia and Herzegovina, the most ethnically mixed of all the Yugoslav republics.

Given the precedent set by Slovenia and Croatia, many could see that Bosnia would soon follow suit and declare independence. Under the pretext that the minority Serb population in Bosnia was being oppressed and needed protection,[31] Serbs in Bosnia were mobilized in anticipation of a Bosnian declaration and the stage was set for the ethnic cleansing that was soon to follow. While much of the violence was directed by the Serb regime in Belgrade, it was the Serb population in Bosnia led by their leader, Radovan Karadžić, that facilitated and made possible the campaign of ethnic cleansing in a newly Serb republic.[32] A number of Bosnian Serb paramilitaries had earlier been formed, armed, and trained with covert support from the Milošević government,[33] and so, when the declaration of Bosnian independence came in March 1992, Bosnian Serb forces, led by these militias, were ready to begin a campaign of violently terrorizing Bosnian Muslims. Sensing an opportunity, Croatia also got into the act in an effort to gain territory in adjacent parts of Bosnia. The fighting in Bosnia was marked by extreme violence, typically against civilians, and included mass murder, extralegal executions, torture, systematic sexual assault and rape, displacement and deportation, and a variety of other war crimes and human rights abuses.[34] It was also notable for the reappearance of concentration camps in Europe in places such as Omarska, Keraterm, and Trnopolje, where Bosnian Muslim men and boys were illegally arrested, detained, tortured, and often murdered. The most infamous of all of the atrocities, however, must be the massacres at Srebrenica, which occurred in July 1995 after the Bosnian Serb Army led by General Ratko Mladić and several militia groups killed more than eight thousand Bosnian Muslim men and boys.[35] It wasn't until late 1995 that the killing was brought to an end after NATO airstrikes forced the Serbs to the negotiating table in Dayton, Ohio, resulting in a peace agreement being signed on November 21, 1995. It was all too late, however, for the approximately one hundred thousand dead and missing, the vast majority of whom were Bosnian Muslims and almost 40 percent of whom were civilians.[36] Examining only Bosnian Muslim casualties, the ratio of civilian victims is even worse, with civilians making up about 81 percent of the victims.

The case of Bosnia and its experience with "ethnic cleansing" clearly illustrates how apparently stable and tolerant societies can degenerate into ethnic violence and genocide given the wrong circumstances and leadership. As I asserted earlier in this chapter, while the former Yugoslavia does not represent an example of a situation in which war and genocide developed directly because of climate change, it nevertheless reveals much about the ways in which political and social instability can create an environment in which unscrupulous political leaders are able to fan and exploit historic prejudices and animosities in order to advance their individual, national, and ethnic agendas. Underlying faults and schisms in Yugoslavia were first exposed by social and political changes sweeping the region and then manipulated by politicians and nationalists who felt simultaneously threatened by the weakening of the Yugoslav federal government and empowered by a resurgence of nationalist identity. In an era of climate change, in which nations around the world will face many unprecedented challenges and pressures, such a scenario in which political and social leaders react to changes, insecurity, and anxiety by exploiting and manipulating preexisting prejudices is all too plausible and perhaps even likely.

The problems of a changing climate will not only exacerbate existing problems and cleavages but also create new ones. Societies already struggling to deal with issues of poverty, social unrest, environmental degradation, political instability, ethnic tension, and all of the other tribulations endemic to many nations and regions around the world, will find these same issues intensified by a rapidly changing world. Issues such as resource loss, population dislocation, flooding, prolonged drought, heat waves, more extreme storms, and loss of crops and arable land will all be added to the mix. The sociologist Christian Parenti refers to this situation as a "catastrophic convergence," which he defines as a "collision of political, economic, and environmental disasters"[37] in which these misfortunes don't just happen sequentially but will occur in concert with each other, each amplifying the impact of the others. The risk is that such an intensification effect will overwhelm the ability of states to cope and adapt. One or two catastrophic changes may potentially be managed; human communities have always faced and usually surmounted both human and natural disasters. But when these shocks keep coming in rapid succession, each magnifying the other, then it is quite likely that many states will simply not have the capacity to cope. Even in those places that are more lightly impacted and/or have the resources, wealth, and political capacity to respond to the demands, the ability to successfully address these challenges can be eroded over time. The speed and magnitude of the changes make such an erosion almost inevitable. Nations and population groups, just like individuals, have a finite ability to change, especially within relatively short periods of time. The political scientist James R. Lee makes this precise point: "Human society is capable of enduring events and seasons, but as these events and seasons accumulate over many years or even decades, accumulated wealth begins to draw down and eventually dissipates. Without renewal of society's wealth, human health and well-being decline, and over time the society itself may collapse."[38]

Moreover, even though wealthy societies will initially prove more resilient to climate shocks, they will fall farther and faster once their breaking point is reached. Modern technological and urban societies depend heavily on highly complex and interdependent systems that are also highly fragile. If one or more pieces of the system fail, it can produce a cascading effect on other systems with each subsequent breakdown multiplying the impact throughout society.[39] As regions, na-

tions, and communities struggle to cope with the myriad problems arising, both directly and indirectly, out of local and regional changes to
their climate, we can identify a number of issues that will play important roles in fostering and facilitating violent solutions and reactions.
While by no means definitive, these should suffice to illustrate some of
the ways in which climate change can impact society and heighten the
risk for violent conflict. As we will see, these pathways to violence
operate on a number of different intertwined and often overlapping
levels, each potentially influencing and amplifying the others and in
turn being influenced and amplified by them. For ease of discussion,
we can sort these conduits to violence into three types: situational,
ideological, and psychological.

Situational factors concern those influences external to any individual or group and which consist of large-scale environmental, institutional,
and structural conditions. Such situational circumstances might, for example, include preexisting conflicts that escalate into genocidal violence. As societal stresses, such as resource loss, ethnic and racial tensions, population dislocation, and similar issues, mount in a given location, small-scale ongoing conflicts may take on a new intensity and
importance as the stakes become more important. In such a situation,
there is a clear risk of the violence escalating and becoming more radicalized: a recipe for the development of the genocidal impulse. Another
situational factor could be preemptive wars waged in order to ward off
potential aggression from neighboring states or from internal enemies.
In such a case, it may be that a state or population group could choose
to proactively attack another if they believe themselves or their resources to be at risk.[40] Such a move to forestall or prevent possible
aggressive action on the part of other groups may be justified through
perceptions of self-defense. Such arguments have often been relied
upon in the past and used to justify wars and genocides throughout
history—self-defense being one of the more universally approved forms
of violence.[41] Another likely situational factor facilitating violent conflict
will be state failure.

The world is primarily organized into nation-states in which a political structure provides public goods and services to all people living
within a geographically bounded region.[42] Within that territory, states
control and manage populations by organizing social, political, and economic life and maintaining law and order within the boundaries of that

land. By tradition and law, states possess the sole legitimate authority, known as sovereignty, over a geographic region, a concept that extends to the belief that the state possesses sole lawful use of violence within the area under its control, which is why British sociologist Anthony Giddens refers to states as "bordered power containers."[43] When states weaken and fail, they typically become prone to lawlessness and various forms of violence since the government cannot assert its authority over its own territory and is unable to effectively govern.[44] In such a situation, as normal services are disrupted and society destabilizes, people are naturally attracted to leaders and movements promising to protect and provide for them. The power vacuum created by a failing state, therefore, opens the door for violent criminality and the rise of alternate nodes of power such as warlords, revolutionary movements, organized crime groups, and terrorist organizations as traditional social institutions, public services, and mechanisms of control are weakened and lost. Weak states can further contribute to conflict by lashing out and using repression and violence to try and reassert control and reassure citizens. A desperate government is often a dangerous one. It's also possible that elite rule can fragment with political factionalism arising and contributing to violence since those challenging those in power "to protect and promote their own interests . . . must destroy or cripple the regime and elites who operate it. Irregular and forcible power seizures, attempted seizures, or a widespread expectation that such seizures may occur are thus a by-product of elite disunity."[45] Failed states are not peaceful places. Estimates suggest that violence and wars in failed states since the 1990s have resulted in around eight and a half million deaths with a further four million people being forced to flee.[46] In the present day, according to the Fragile States Index, South Sudan, Somalia, the Central African Republic, and the Sudan are the most fragile/failed states, followed closely by the Democratic Republic of the Congo, Chad, Yemen, Syria, Afghanistan, Guinea, Haiti, Iraq, Pakistan, Nigeria, Côte d'Ivoire, and Zimbabwe.[47] Most of these are violent and dangerous places, whether through rampant criminal violence, civil war, warlordism, terrorism, or some combination of these. Thankfully, not all states are equally at risk for failure.

States fail for many reasons, including their specific circumstances, history, geography, economics, political structure, culture, and leadership. One comprehensive analysis of state failure found four main indicators closely connected to state failure:

- Quality of life, that is, the material well-being of a country's citizens.
- Regime type, that is, the character of a country's political institutions.
- International influences, including openness to trade, memberships in regional organizations, and violent conflicts in neighboring countries.
- The ethnic or religious composition of a country's population or leadership.[48]

These factors proved highly consistent in predicting historic cases of state failure and tell us that poorer nations that are divided along ethnic or religious lines and have low levels of development and weak democracies or authoritarian regimes, are the most vulnerable for failure. This is especially the case if they are located in regions where violence is taking place in neighboring states, since such conflicts often serve to destabilize entire regions and spill across national borders. Furthermore, this analysis looked at state failure that had degenerated into genocide and found that the odds greatly increased when social elites came from one ethnic group and where national ideologies were exclusionary in terms of defining who belonged and who didn't.[49]

Another useful way to examine the issue of state failure is to consider the notion of resilience. Most often used in studies of ecological systems,[50] and more recently in regards to bullying and school shooters,[51] resiliency refers to the ability to absorb changes and the speed with which systems, structures, and/or individuals are able to bounce back from shocks and setbacks.[52] The greater their resiliency, the better they are able to weather adversity and challenge. Resiliency can also be applied to understanding how human communities, societies, and governments respond to hardship and stress. The less resiliency a state possesses, the more vulnerable it is to failure. Three main variables appear to determine such an eventuality:

1. Extent of climate change impacts
2. Sensitivity to those effects
3. Ability and willingness to adapt to the new reality[53]

In other words, states vulnerable to failure are those in locations where the changes wrought by climate change are significant and have a strong effect on social institutions and structures. Much will then depend on the ability of that society's institutions and leaders to confront the situation. In a word, this is about resiliency. Digging deeper, we can suggest that vulnerable states will most often share a number of important characteristics. First and foremost, they will tend to be poor nations. Wealth is strongly correlated to the resiliency a nation typically enjoys. Ironically, poor nations are often home to a great deal of natural resources but remain stuck in extreme poverty because they rely on extractive economies that revolve around exploiting resources for export. Commonly referred to as "the resource curse," it is a sad paradox that nations rich in natural resources are often more poor and less well governed than countries without an abundance of such assets.[54] Reliant on extractive economic systems, such countries are prone to corruption and are characterized by a small number of social elites who've enriched themselves at the expense of the vast majority of the population.[55] They also do not invest in constructing stable infrastructures, creating economic or educational opportunities for the population at large, or building social and economic institutions. Nor do they create cultures that respect the rule of law. Yet without such investments, the multiple impacts of climate change will be difficult to address and remedy. Without a strong public health system, for example, outbreaks of disease cannot easily be confronted and may increase social instability and unrest. Disease vectors in many parts of the world are expected to change since a warmer world will increase outbreaks of disease such as malaria, dengue fever, and cholera and allow water- and insect-borne diseases to spread into regions once free of such infectious pathogens. Similarly, where food and water scarcity exists, the lack of a reliable transportation infrastructure, such as good roads and railways, will make it difficult to redistribute needed supplies from unaffected areas to those regions experiencing shortages. These kinds of failures can have important consequences for social, political, and economic stability. Marked by a distinct lack of public services in areas such as transportation, energy, health care, and education, these states are prone to high levels of corruption, widespread criminality, and collective forms of violence.

Weak or failing states lose the ability to govern and deliver crucial services to their citizenry and are characterized by a lack of state authority and control, especially in more remote and peripheral regions of the country. The weaker they become, the less they are able to generate the capital and capacity needed to effectively confront mounting challenges. Val Percival and Thomas Homer-Dixon succinctly capture this dynamic:

> Falling agricultural production, migration to urban areas, and economic contraction in regions severely affected by scarcity often produce hardship, and this hardship increases demands on the state. At the same time, scarcity can interfere with state revenue streams by reducing economic productivity and therefore taxes. . . . Environmental scarcity therefore increases society's demands on the state while decreasing [the state's] ability to meet those demands.[56]

Not surprisingly, such a situation is highly susceptible to the development of collective violence since the failure of a state to meet the needs of its people and provide for public safety and security results in a loss of legitimacy in the eyes of its population, which increasingly must fend for itself and confront rising levels of crime, insecurity, and material deprivation. Often alternative centers of power, such as warlords, criminal cartels, and revolutionary movements, arise in the vacuum that results from state weakening and failure. Political scientist Robert Rotberg describes this process as follows:

> In the last phase of failure, the state's legitimacy crumbles. Lacking meaningful or realistic democratic means of redress, protesters take to the streets or mobilize along ethnic, religious, or linguistic lines. Because small arms and even more formidable weapons are cheap and easy to find, because historical grievances are readily remembered or manufactured, and because the spoils of separation, autonomy, or a total takeover are attractive, the potential for violent conflict grows exponentially as the state's power and legitimacy recedes.[57]

Rotberg is not alone in making such an assessment. In his work on state failure in Africa, Robert Bates examines a number of cases where the government collapsed and the society descended into widespread anarchy and collective violence.[58] Bates's analysis, similarly to Rotberg's, identifies political leadership as a key factor in state failure. Specifically, he argues that state failure in Africa resulted from political leaders failing to invest in civil society, infrastructure, rule of law, and economic development. Instead, they allowed corruption, mismanagement, authoritarianism, and intolerance to flourish, which, in turn, led to more competition and conflict over resources and a citizenry increasingly turning to movements, rebels, and warlords offering protection in exchange for allegiance and loyalty. Interestingly enough, most of the nations analyzed by Bates were home to many ethnic groups, but importantly, such heterogeneity did not cause violence, even though such conflict developed. Ethnic conflict did not *cause* state failure, but was instead a *result* of it.[59] The collapse of state authority and control allowed ethnic hostility to flourish, which then devolved into violent struggle. In many African nations, ethnic groups occupy distinct territories, which meant that attempts to distribute and redistribute resources from one part of a country to another often took on ethnic dimensions, or as Bates asserts, "Ethnic conflict is not a 'clash of cultures,' then, but rather a struggle over the regional allocation of resources."[60] Clearly, in addition to state failure, resources are also at the heart of many conflicts.

Simply put, the word "resource" refers to supplies and stockpiles of materials, wealth, people, and other goods that allow people and communities to survive, function, and produce needed services. Ore, rare earth metals, wood, oil, gas, water, land, food, people, and countless other things can all be defined as resources since they can be drawn upon, harvested, utilized, transformed, exchanged, and consumed for individual and community needs. Resources are what allow a state to meet the basic survival needs of its citizens, including providing food and water, maintaining a stable government and infrastructure, providing housing and energy, and implementing policies to combat or at least cope with climate change impacts. One of the first to note this link was political scientist Thomas Homer-Dixon, who argued that environmental scarcity can bring about a social and political deterioration resulting in violent conflict.[61] In a nutshell, his argument focuses on what he terms "resource capture," and "ecological marginalization," with the first phrase referring to ever scarce resources leading one group to try and take over remaining supplies from another population, while the latter term concerns unequal access to resources such as land and water creating pressure for migration into less sustainable areas. Both mechanisms can reinforce each other and, especially when combined with population growth, lead to violent clashes and conflict.[62] In the modern era, resources are already flashpoints around which conflicts have developed and one has only to look at world events to see how competition, conflict, and war have frequently arisen out of disputes over vital resources such as oil and water.[63] Often these struggles have become intertwined with identity issues such as ethnicity, nationalism, and religion, but these overt dynamics, important as they are in their own right, mask the underlying resource dynamics.

We may find that under the pressures and stresses of climate change, states will be more willing to engage in war to either protect resources or acquire new ones, such as occurred when Iraq invaded Kuwait in 1990. This invasion occurred because of Kuwaiti control of the vast Rumaila oil field reserves and Iraqi dictator Saddam Hussein wanted those valuable resources for Iraq.[64] Failing such direct methods, some states may instead choose to indirectly support proxy forces in order to accomplish the same ends. Such tactics can be facilitated by state failure in the targeted nation and the resulting anarchy, lawlessness, and criminal violence resulting from the power vacuum. It is terribly ironic that nations have sometimes benefited from and fostered state failure and the resulting chaos in neighboring countries. A good example of this concerns the Democratic Republic of the Congo (DRC), formerly known as Zaire until it was renamed in 1997.

In April 1994 genocide broke out in Rwanda during a civil war. The Tutsi minority population, long stigmatized and scapegoated by Hutu extremists, were targeted for elimination by Hutu militia groups, police and military personnel, and mobs of ordinary citizens.[65] Many local perpetrators were motivated in part by the chance to loot the property and acquire the land of their Tutsi neighbors. Rwanda is a densely populated country with many subsistence farmers existing on very small patches of land. In 1994, about 95 percent of the population was rural and 80 percent of Rwandan farmers had less than two hectares of land,[66] so such incentives were powerful inducements for participation. Even at the individual level, resources can play a part in motivating killers during genocide.[67] This isn't to suggest that resources were the only reason why so many participated in the killing, but to ignore the motivation of material gain is to disregard an important piece of the puzzle as to why so many joined in and took part.

As the Hutu-led government began to lose the war and fearing retribution for the genocide they had inflicted on their Tutsi fellow citizens, many Hutu fled into Zaire, including soldiers, various local and national politicians, ordinary civilians, and even members of militia groups.[68] Safely ensconced in Zaire, the genocidal paramilitaries reorganized, took control of the refugee camps, and began plotting their return to Rwanda. The refugee camps were soon rife with violence as some of the former perpetrators began victimizing their fellow refugees. They also began to organize and carry out raids back across the border into Rwandan territory. The situation got so bad that Médecins Sans Frontières packed up and left the camps.[69]

Zaire at this time had a highly corrupt and weak government run by the dictator Mobutu Sese Seko. His was the classic "resource curse" dictatorship. Notorious for enjoying a profligate lifestyle while the vast majority of his people lived at a subsistence level, Mobutu gave his nominal support to the Hutu fighters who continued to attack, pillage, and murder on a wide scale in their raids across the border.[70] Enough was finally enough, and in 1996 the government of Rwanda responded to the cross-border raids by sending in their military and, together with Zairean rebels and Ugandan troops, they helped overthrow the Mobutu government. Soon the country was a destabilized mess without a strong central government and from that time to the present, violence has wracked this region with local and regional militia groups fighting each other and the forces of other nations. Not only has Rwanda been a major player in the violence, but Uganda, Angola, Burundi, and Zimbabwe have also been heavily involved at times. Estimates suggest that the wars and anarchic violence have killed up to 5.4 million, according to one educated guess, and internally displaced another million.[71] One particularly notable quality of the wars, rebellions, insurgencies, and genocide that have ravaged this region has been the extremely widespread targeting of women for sexual assault, leading one United Nations officer to comment that in the Congo it was more dangerous to be a woman than it was to be a soldier, while a reporter labeled the eastern Congo as the rape capital of the world.[72]

In trying to explain the extreme levels of violence in the DRC, we need to acknowledge that state failure, power politics, and ethnic hostility played important roles in facilitating and shaping the wars and genocidal assaults on different tribes and population groups in this region.

Underneath it all, however, were natural resources. One of the poorest nations on earth in terms of standard of living and per capita income, the Congo is home to some of the world's richest deposits of gold, copper, diamonds, and lumber, not to mention farmland and hydro-power.[73] Ironically, it has been the tremendous reserves of natural re-sources that have helped perpetuate the conflict. Various combatants, often with the involvement and support of regional powers and multina-tional corporations, funded campaigns of violence and genocide through the exploitation of these abundant and valuable resources. Rwanda, for example, while initially concerned with ending the vio-lence emanating from the refugee camps, became increasingly preoccu-pied with harvesting the wealth of this troubled and vulnerable coun-try.[74] Protected by Rwandan troops, encampments and facilities were set up to facilitate the extraction and removal of valuable resources, such as tin, copper, and coltan: an ore containing minerals vital for many high-tech products such as cell phones. Rwandan military forces were also deployed to guard those shipments from the war zone back to Rwanda.[75] The Ugandan government and military have been similarly implicated in such resource exploitation.[76]

The case of the DRC certainly illustrates the ways in which war and genocide can revolve around resources, as well as the ways in which states can directly and indirectly exploit a situation for their own bene-fit. As climate change increasingly impacts our world, resource con-cerns are likely to take center stage in many locations, especially since many important resources are already close to exhaustion.[77] In many places, something as basic as providing enough food for a population will present challenges as drought and flooding decrease agricultural productivity. Even land itself, especially arable land, will be seen as a resource worth fighting and killing for. In some parts of Ethiopia, local subsistence farmers are already finding steel fences erected around vast swaths of land in their communities. These segregated tracts of farm-land have been bought by Saudi Arabia as part of a plan to buy arable land overseas and grow crops for export back to Saudi Arabia.[78] How will those local farmers react when the struggle to meet their own need for food intensifies? Violence has already broken out as Ethiopian po-lice and military personnel have cracked down on local protests orga-nized by local villagers and farmers.[79] Such repression and the social unrest attending land grabs will only deepen as food security becomes ever more tenuous in states struggling with shortages. Land is not the only resource around which such conflict is already playing out. China, for example, has dramatically increased its investments in extracting oil from the Sudan and South Sudan,[80] and has been intensifying its claims in a territorial dispute over the Spratly Islands in the South China Sea because of potential oil and gas reserves.[81] Such attempts to acquire needed resources will only intensify if scholars, such as Michael Klare, are correct. Presenting a bleak assessment for resource-based conflict, he suggests that "given the growing importance ascribed to economic vigor in the security policy of states, the rising worldwide demand for resources, the likelihood of significant shortages, and the existence of numerous ownership disputes, the incidence of conflict over vital materials is sure to grow."[82]

Klare further asserts:

> For nation-states, the fight for resources has equally high stakes; those that retain access to adequate supplies of critical materials will flourish, while those unable to do so will experience hardship and decline. The competition among the various powers, therefore, will be ruthless, unrelenting, and severe. Every key player in the race for what's left will do whatever it can to advance its own position while striving without mercy to eliminate or subdue all the others.[83]

This is indeed a Hobbesian nightmare vision of the future in which a dog-eat-dog mentality dominates international relations. In summary, a variety of structural issues arising out of or exacerbated by climate change will facilitate various social, economic, and political conditions conducive to the development of various forms of collective violence. Things such as preemptive wars, the amplification of preexisting conflicts, state failure, and aggression to either protect existing resources or acquire new ones are all possible pathways fostering violent outcomes, solutions, and reactions to the challenges of climate change. Importantly, however, we must acknowledge that such conditions, in and of themselves, may not necessarily spark violent conflict, especially not wars and genocide. Individuals and communities do not react simply or solely out of materialistic needs and pressures. Instead, people also need to be motivated by a variety of beliefs and values. We are, after all, symbolic beings for whom such concepts and perceptions play important roles in shaping action and behavior.

In *Making and Unmaking Nations: War, Leadership, and Genocide in Modern Africa*, political scientist and genocide scholar Scott Straus conducted a case study of a number of African nations that were embroiled in conflict and attempted to identify why some of those wars escalated into outright genocide while others did not.[84] Straus found that the two most important determinants were what he termed "founding narratives" and the political leadership in power during the conflict. Founding narratives, according to Straus, are the belief systems that define the values and core identity of a political and national community. These founding narratives are instrumental in shaping and influencing the dominant political, national, and cultural identities within a society and are thus highly influential in guiding decision-making processes. Belief and identity often guide action.[85] Straus found that when perceptions of threat and crisis were heightened, such as during violent conflicts, states with exclusionary founding narratives were much more likely to resort to genocidal violence against outsider and minority groups. On the other hand, in those embattled nations where the founding narratives were more inclusive and tolerant or where counternarratives were also present, outbreaks of genocidal violence did not occur.

Straus's analysis highlights the role of political leadership and ideology in either facilitating or hindering collective violence. We need to remember that genocides represent rational efforts to realize certain objectives, but it is a rationality guided and shaped by nonrational belief systems. To some extent, all genocides have an underlying ideological element that frames the policies and practices of persecution and destruction in ways that justify and legitimize them. The genocidal impulse needs large segments of society either passively accepting or actively participating in the scapegoating, marginalization, persecution, and destruction of minority groups. Ideology is one powerful tool for facilitating such widespread acceptance. It helps shape belief in the virtue of such violent maltreatment and allows perpetrators to understand their destructive actions as constituting what sociologist Jack Katz refers to as "righteous slaughter."[86] Some ideologies, for example, can suggest that participation is a necessary obligation for country, people, race, or ethnicity.[87] Principal among such belief systems is nationalism, which holds that a group shares a common culture, history, and tradition that connects members with a national community and distinguishes them from people outside of that community.[88] Nationalism also stresses certain qualities such as loyalty, obedience, duty, and allegiance. Patriotism, in other words.[89] While nationalism as an ideology can unify and strengthen a sense of belonging and community, it also has a dark side, since by emphasizing the in-group it also highlights an out-group or as Vamik Volkan puts it, "While the idea of nationalism may be linked to liberty and universalistic ideals, it also sometimes led to particularism, racism, totalitarianism, and destruction."[90] Noting the same tendency, another scholar simply refers to nationalism as "ideologies of exclusion."[91]

Nationalism and the social and political divisions such ideology engenders will have important ramifications for war and genocide as nations confront a changing climate. As nations deal with resource loss, environmental degradation, population dislocation, and various other climate-induced problems, violence against groups defined as outsiders, as posing a threat, or as being in possession of needed resources may be framed by nationalistic ideologies as a patriotic duty. The legitimacy or illegitimacy of violence lies in how that act is defined, not in any intrinsic quality of the act itself, and ideology can powerfully shape those perceptions so that those participating in genocide and other collective forms of violence come to accept the morality and necessity of their destructive actions. What is ironic is that perpetrators of genocide continue to define themselves as good, moral, and patriotic citizens.[92]

Nationalistic beliefs are often supported by various dehumanizing and denigratory portrayals of minority and out-group populations and these expedite and make persecution easier. Dehumanization, or pseudospeciation as it is sometimes referred to, involves defining the targeted group as a lesser form of life and has the effect of removing those individuals from what some have termed a "moral circle,"[93] and others have referred to as the "universe of obligation."[94] The psychologist Albert Bandura developed the theory of moral disengagement to show that individuals learn to behave in ways that do not challenge their personal values and morality because that would result in self-condemnation and self-criticism.[95] Resourceful as always, people have crafted ways of selectively disengaging their internal moral prohibitions against violent and destructive behavior in order to avoid seeing themselves in a bad light. Violence, for example, can be reframed as being morally sanctioned and justified. Such moral disengagement is facilitated through dehumanization of the intended victims. It is much easier to remove ethical restrictions against violence when the victims are defined as being less than us or nonhuman. During genocides, for example, the victim groups may be labeled as cockroaches, insects, and vermin, which serves to make the violence against them easier and more palatable.[96] The language of genocide is rife with such dehumanizing language, facilitating the persecution and destruction of identified populations. In fact, some scholars suggest that dehumanization is one of the most important preconditions needed for genocide.[97] Dehumanization, in other words, defines those people as existing beyond the range of our civility, compassion, care, and concern. Moral and ethical standards and guidelines do not apply to groups so excluded. Such beliefs are particularly dangerous as they remove normative obstacles to harming those so defined. Not only do the perpetrators of genocidal violence continue to define themselves as moral actors, but they also define their victims in such a way so as to not activate their moral responses. It is much less difficult to hurt or kill those who are not only different from you, but also less than you. Such ways of thinking about other groups serve to widen the social distance between the victims and the perpetrators and thus enable victimizing them. Nationalism and dehumanization, therefore, are particularly potent ideological tools enabling violence and persecution.

Dangerous ideologies also often highlight historic victimization and loss and are particularly effective since they reinforce and perpetuate a group identity as having been wronged and consequently provide a ready-made justification for genocide and other forms of violence. This is what Vamik Volkan refers to as a chosen trauma, in which a group identity is built around a past grievance that can be utilized to justify persecution and violence in the present.[98] The political scientist and genocide scholar Herb Hirsch describes this process when he points out that "nation-states in particular use, create, or respond to myths about themselves that they wish to perpetuate, and, in turn, the myths are used to justify or rationalize policies that the leadership of the state wishes to pursue."[99] While belief systems can shape circumstances, circumstances can also shape belief systems, and this is where psychology, our last pathway to conflict, comes into play.

In trying to explain the onset of genocide, scholars have long identified tough times as one important factor leading up to the genocidal impulse because of the negative attitudes and opinions that they foster. Ervin Staub, for example, asserts that difficult life conditions can threaten people's lives, sense of security, well-being, self-concept, and worldview, and such threats can activate deep-seated psychological reactions that make hostility, aggression, and violence more likely.[100] Staub further emphasizes that to understand why states target minority groups for elimination one must understand how circumstances shape support for policies of destruction. As he puts it:

> Why does a government or a dominant group turn against a subgroup or society? Usually difficult life conditions, persistent life problems in a society are an important starting point. They include economic problems such as extreme inflation, or depression and unemployment, political conflict and violence, war, a decline in the power, prestige, and importance of a nation, usually with attendant economic and political problems, and the chaos and social disorganization these often entail.[101]

His theoretical explanation has a fair amount of empirical support backing it up. Criminological research, for example, has found that groups often develop much more reactive and punitive attitudes during uncertain economic times, periods of high crime rates, and eras of social and cultural change.[102] People find such situations unsettling, disorienting, and fear inducing. As a result, they react in protective, defensive, and often aggressive ways. The sociologist Randall Collins describes such a context as "confrontational tension," in which a buildup of tension and fear develops over time and can result in a violent release of these emotions against individuals or groups that are seen as representing a threat.[103] This explosion of violence can represent a "forward panic," a term Collins uses to characterize the sudden release of fear and tension through violent actions. Examining the dramatic increase in popular support for more punitive measures and the implementation of such policies in both the United States and Great Britain during the late 1970s through the 1980s, British sociologist David Garland suggested that widespread economic and social changes led to a general increase in feelings of insecurity and a deeper sense of physical risk and fear.[104] In such a social, economic, and political climate, harsher and more radical measures and attitudes became more acceptable and resulted in widespread support for more extreme policies. Difficult and stressful

life conditions engendered frustration, uncertainty, and fear resulting in
an increase in hostility and anger toward groups defined as dangerous;
in this particular case, that meant criminals. A similar process marked
the Nazi rise to power, which was aided and abetted by the harsh terms
of the Versailles treaty ending World War I, the postwar violence and
revolutions common throughout Germany during the 1920s, and the
Great Depression. Times were tough and many ordinary Germans were
attracted to the message of the Nazi Party that blamed the Jews for all
of Germany's misfortunes. The reason why it worked was because Europe had a long tradition of anti-Semitism that had often involved the
scapegoating of Jews for many historic problems.[105] In fact, Hitler himself once said, "Experience teaches us that after every catastrophe a
scapegoat is found."[106] The Nazis also promised to make Germany safe
again and to return it to its historic luster. All of these messages resonat-
ed with a population eager for such reassurances. Analogously, the social, political, and economic changes transforming Yugoslavia in the late
1980s and early 1990s provoked widespread unease and fear and helped
propel the rise of radical politicians promoting policies of ethnic chau-
vinism that ultimately fractured the nation and unleashed the policy of
ethnic cleansing in Croatia and Bosnia.

In an era of climate stressors, it's easy to imagine similar processes
and sentiments targeting not only criminals, but also recent immigrants,
those from different religious, racial, and ethnic groups, those perceived as benefiting unfairly from a society, those who are not seen as
contributing to the common good, or those who have simply been victimized in the past. The frustration-aggression hypothesis teaches us that
when people are blocked from achieving an expected goal, they often
become frustrated and such feelings can easily turn to anger.[107] Such
frustrations can result from people believing that their economic opportunities and jobs are being blocked by others, or that their safety is
being compromised by individuals or groups who are on the social or
political margins of society. Think of the backlash against immigrants in
recent years in many parts of the world. Fueled in large part by the
economic uncertainty and hardship experienced after the global economic downturn in 2008 and by fears of Islamic terrorism in the aftermath of a number of well-publicized attacks in western Europe, anti-
immigrant sentiment and legislation have surged and popular narratives
have increasingly painted immigrants as posing an existential threat to

traditional ways of life, the host nation, and even Western civilization itself. Almost certainly as a consequence of such rhetoric, many nations have seen an increase in right-wing extremist and nationalist political movements often accompanied by violent protests and action.[108] This isn't a new phenomenon. One recent study looked at eight hundred elections spanning more than 140 years and found that in the wake of a financial crisis, nativist and nationalist politicians and movements typically saw a dramatic upswing in their share of the vote and, importantly, also found that much of their success was due to a willingness to blame and scapegoat minority populations and foreigners.[109] It is quite likely that such hostility against minority groups will only intensify and result in more punitive policies, and in some cases, repression, discrimination, and violence, given that the number of refugees is expected to increase dramatically in the coming years.[110]

This is especially possible in situations where resources, jobs, physical security, and public services become scarce and competition intensifies. In such locations, it's quite possible for popular sentiment and anger to turn against those defined as outsiders, dangerous, and superfluous. Individual life will be seen as being less valuable because of too few jobs, too few resources, and too few services for too many people.[111] The frustration resulting from blocked opportunities, a loss of stability, and perceptions of increased risk can easily morph into anger. This is a recipe for confrontation and violence because communities and population groups may choose to target migrants and minority groups for persecution and elimination. We know that during times of crisis and uncertainty, states often become more authoritarian and intolerant of dissent, a fact that also contributes to a higher risk of persecution.[112] Furthermore, in those places where the state is weakening or failing, the government may resort to more punitive measures and violent crackdowns in an effort to reassert control and reassure its citizens that it has the matter in hand. Without putting too fine a point on it, this is how collective violence, from pogroms to war to genocide, becomes not only possible but probable.

Political and social leaders have often capitalized on frustration, anger, and prejudice among a population in order to gain or retain power and scapegoat select groups. Scapegoating may be done in order to sidetrack attention away from failed policies or to provide populations with easily understood answers that resonate emotionally because they build on preexisting stereotypes and prejudices. Complex situations do not always present easily understood or satisfying answers and scapegoating a vulnerable group can appeal because it provides answers that are quickly and easily grasped and can provide a focus for anger and frustration. Sometimes political, military, and/or religious leaders may rely on scapegoating in order to deflect responsibility away from themselves and their policies. In the same way that the Nazi Party scapegoated the Jews for Germany's defeat in World War II and for the collapse of the German economy during the Great Depression, political leaders and population groups dealing with difficult circumstances may find it easier to attribute responsibility to vulnerable groups for the problems of a particular place and time than it is to acknowledge their own failings or inability to solve intractable problems.

We also shouldn't ignore the fact that political leaders are themselves often prone to the same prejudices, fears, and nationalist mythologies that their fellow citizens share. During stressful times, these sentiments can easily shape policy decisions. Jerrold Post, in summarizing a body of research on decision making during crisis, points out that many leaders may have their decision-making skills compromised when under stress because they often

1. Develop a more fixed mind-set and thus lose creativity and flexibility. Violence, a straightforward solution, may thus appear more attractive.
2. Narrow their awareness and focus so that they only perceive a limited range of options.
3. View the present in terms of the past; a dangerous situation if a history of conflict and intolerance mark that society.
4. Perceive the situation according to familiar mental models or scripts such as those that portray a stigmatized group in stereotyped and prejudicial ways.
5. Are more likely to fall prey to attribution bias. That is, they are more likely to see the actions of others as being guided by internal psychological processes such as malevolence and hatred, rather than seeing actions as being brought about by impersonal and external forces.[113]

In other words, during times of high stress and crisis, such as those brought about by the projected impacts of climate change, leaders and other decision makers can suffer from impaired decision-making processes and such a deterioration may facilitate the development of violent solutions targeting vulnerable populations. When combined with the other factors discussed above, such compromised thinking and judgments may well have lethal consequences.

In summary, I've suggested that while climate change will not directly cause wars and genocides, it will certainly create conditions and mind-sets conducive to the development of intergroup hostility, tension, and violent conflict. These, in turn, will help enable ethnic and communal violence, wars, and genocide. Certain structural circumstances in particular will contribute to this heightened risk. These include ongoing or preexisting conflicts escalating into genocidal violence; preemptive wars initiated as defensive measures; states weakening and failing thus allowing for the rise of criminal violence, warlords, and revolutionary movements; and conflicts arising out of the desire to protect scarce and valuable resources or to acquire new ones. Psychologically speaking, I've also suggested that populations struggling with the realities of a changed environment and all of the structural problems arising out of those altered circumstances will be more receptive to viewing certain populations as outsiders, dangerous, and responsible for many of the difficulties being experienced by their communities. Consequently, they will be more likely to support ever more radical policies targeting those groups and to participate in scapegoating and persecution. The risk of such action will be especially high in those nations and regions where certain kinds of dangerous ideological systems are prevalent that can be used and manipulated by social, political, and religious leaders. This isn't to suggest that conflict, war, and genocide are inevitable in such places, but rather that the possibility of such outcomes will be heightened. At this point, it may be useful to more deeply explore some specific impacts of climate change in order to more closely trace connections between such challenges and war and genocide. A good starting point for such a discussion is to focus on one extremely important resource that will be dramatically impacted by climate change, and that is water.

4

WATER, VIOLENT CONFLICT, AND GENOCIDE

The wars of the twenty-first century will be fought over water.
—Ismail Serageldin[1]

Of all the resources that we rely on for survival in today's world, water is the least appreciated and certainly the most misunderstood.
—Brian Fagan[2]

While not an obvious issue to us in twenty-first-century America, management of drinking water as a resource—who gets it, when they get it, and how much they get—has been a life-and-death matter for much of human history.
—James Salzman[3]

On December 17, 2010, a young man named Tarek el-Tayeb Mohammed Bouazizi doused himself in gasoline on a street in front of the governor's office in the Tunisian town of Sidi Bouzid and set himself on fire.[4] What could possibly drive a young man to such a desperate act? It turns out that he killed himself as a protest against the corruption and harassment he had been experiencing from local government officials. A street vendor who sold fruit to support himself and his family, Bouazizi had been slapped and beaten when he had tried to prevent his fruit from being confiscated by government officials.[5] To add insult to injury, one of the officials was a woman who reportedly slapped him in the face, spit at him, took his scales, and insulted his father.[6] The fact that it was a woman who so publically humiliated him added to the shame of this particular confrontation, but this wasn't the first time that he had been mistreated by public officials. He had long been harassed and beaten and suffered the indignity of having his goods confiscated by police officers and other government officials because he didn't have enough money for the bribes that were endemic in his country.[7]

Needing his confiscated electronic scales to sell his produce, Bouazizi went to the municipal building to complain and retrieve his scales, but since he didn't have the necessary funds to bribe officials, he was again beaten and denied the return of his scales. The governor also refused to see him, even after Bouazizi reportedly yelled, "If you don't see me, I'll burn myself."[8] He was a young man who had dropped out of school to support his family and he was keenly aware that they depended on his income for their survival. Humiliated, angry, frustrated, and desperate, he reacted by getting some gasoline from a local gas station, returned to the street in front of the government building, doused himself in gasoline, and set himself on fire. Onlookers quickly tried to douse the flames, but succeeded only after he had been burned over 90 percent of his body. Never regaining consciousness, Bouazizi lingered in the hospital until he died of his burns eighteen days later. In all likelihood, he never would have guessed at the reaction his public self-immolation would generate, but as word spread, protests erupted throughout Tunisia in the hours and days after his death that eventually ousted the authoritarian president of Tunisia, Zine El Abidine Ben Ali, and from there spread across much of the Arab world in North Africa and the Middle East.[9]

While the suicide of Bouazizi was the specific trigger, at a deeper level the wider unrest was the result of long-term economic stagnation, lack of opportunity, political corruption, and authoritarian rule. In a number of cases, most notably Tunisia, Libya, Egypt, and Yemen, the entrenched political leadership was ousted,[10] while in other cases the protests led to increased repression. In one case, Syria, the protests escalated into a full-blown civil war that helped generate the rise of the Islamic State in Syria (ISIS), and which, as of the writing of this book, continues to rage years later. While at first blush it might appear as if this introductory discussion on the Arab Spring and Syria has little relevance to the goals of this book in general, or the focus of this chapter in particular, the truth is that these examples are very salient to the issue of climate change, conflict, and genocide. This chapter is all about water and tracing the ways in which water issues have facilitated the development of violent conflict and genocide in the past, the present, and quite possibly the future. As we will see, water as a resource has had powerful, if sometimes hidden, consequences on human communities and has, on occasion, contributed tremendously to rioting, war, and genocide. The example of Syria in the aftermath of the Bouazizi self-immolation and the subsequent Arab Spring certainly speaks to this influence.

When discontent broke out in Syria early in 2011 as part of the wider social and political unrest sweeping the Arab world, what started out as mass marches and protests quickly descended into fighting as the regime of President Bashar al-Assad responded with harsh measures that rather than quelling dissent, simply hardened resistance against the government and resulted in an armed rebellion that over time took on increasingly sectarian overtones; most of the government forces consisting of Shia Muslims, while the opposition is largely Sunni.[11] Importantly, Assad belongs to the Alawite minority, which only served to intensify the religious dimensions of the fighting. Over time the conflict has evolved into a complex civil war involving multiple opposition groups including Kurds, Hezbollah, and most infamously, the Islamic State or ISIS. These groups have a variety of goals and have fought not only the Assad regime but each other as well.[12] The fighting itself has been especially vicious and has involved numerous war crimes including torture, mass executions, systematic rape, and the use of chemical weapons. Since the beginning of the conflict more than four and a half million Syrians have fled the violence and become refugees, while another six and a half million have been internally displaced.[13] This struggle has sparked worldwide condemnation and outrage because of the many atrocities and has also destabilized the entire region and sucked in a variety of nations, most notably the United States, Russia, and Turkey. While many factors conspired to bring about this horrific situation, such as the authoritarian nature of the Assad regime, a lack of economic opportunity, and the desire for change among many young Syrians, much less well known is that underneath these specific triggers, a deeper source of the conflict has to do with water scarcity.

Between 2006 and 2011, the region was in the grip of a severe drought that hit Syrian society hard, especially in the agricultural and livestock sectors.[14] Food crops, such as wheat and barley, were decimated by the dry conditions with the yield for wheat decreasing about 47 percent, while barley saw a 67 percent reduction.[15] Livestock were also severely impacted, with a decrease in numbers from around twenty-one million animals to around fifteen million.[16] Given that around 40 percent of Syria's workforce was in agriculture, this proved a severe hardship. This was especially true for small-scale farmers and herders.[17] In total, the United Nations estimates that between two and three million Syrians were negatively affected by the drought with many falling further into poverty and food insecurity; about a million and a half Syrians lost their livelihoods and migrated from the rural regions of the country into the cities where many of them found that they had simply traded rural poverty for urban destitution. Unfortunately, the government response to the drought left a lot to be desired since official policy served largely to exacerbate existing social tensions and inequalities and did not significantly ameliorate the situation for those most impacted.[18] Ultimately, when the Arab Spring swept through the region, large numbers of unemployed and displaced Syrians, many of them young men, found a focus and an outlet for their anger that allowed them to channel their resentments and frustrations into a meaningful cause and thus the uprising in Syria began. The scholar of mass movements, Eric Hoffer, once wrote, "It is a truism that many who join a revolutionary movement are attracted by the prospect of sudden and spectacular change in their conditions of life,"[19] and that certainly could describe many of the young people who flocked to the call that promised to overthrow the government that had so clearly failed them. Ironically, those who joined in the protest hoping for an end to a corrupt dictatorship, more freedom, and economic opportunity probably never envisioned the destruction and violence that their actions helped unleash. This is not to suggest that water was the only or even the primary factor leading to the conflict, but it certainly played an important role in creating a situation ripe for conflict, social unrest, and revolutionary action. Water scarcity, in short, helped bring about the violence.

When we think of precious resources, we tend to focus on rare metals such as gold and silver. Perhaps gemstones—diamonds, rubies, or opals—also come to mind. Rarely, however, do we think of water as a resource, much less a precious one. This is especially the case in the industrialized nations of the world where cheap and apparently endless supplies of freshwater are readily available at the turn of a tap. We are lavish, even reckless, in our use of this seemingly inexhaustible resource. We water our lawns, take long, hot baths, fill swimming pools and hot tubs, wash our cars, and build and maintain golf courses in the desert. This extravagant usage is often encouraged by the low cost of water in many developed nations, such as in the United States where we pay less for water than we do for any other utility. We are profligate in our usage and never really consider the true value of this precious commodity. Yet that is precisely what water is: a precious and scarce commodity. In *Poor Richard's Almanac*, Benjamin Franklin captured this fundamental truth when he wrote, "When the well's dry, we know the worth of water," and his words of wisdom may well resonate in the coming years.[20]

All life on earth depends on water and most likely began in a primordial soup of water and chemical compounds about four billion years ago. Plants, animals, bacteria, and viruses all depend on water to survive. For humans, the situation is no different. Human beings require a constant supply of water. Our bodies use it to regulate temperature, flush waste, lubricate joints, allow individual cells to function, and manufacture hormones and neurotransmitters. We are largely made up of water with approximately 60 percent of our bodies being composed of the stuff.[21] The brain and heart are about 73 percent water, the lungs 83 percent, our skin 54 percent, our muscles and kidneys 79 percent, and even our bones are about 31 percent water.[22] These percentages increase and decrease every day depending on how hydrated we are since we lose water constantly through respiration, sweating, and urination and if not steadily replaced the total volume of fluid in the body can fall to dangerous levels. An individual who is active while it is hot outside can require up to sixteen liters of water a day to replenish what is lost,[23] and whereas humans can live up to three weeks without food, we last less than a week without water depending upon our age, health, amount of physical activity, and the environmental conditions. We simply cannot survive without this absolutely essential liquid.

Water is necessary not just for individual human life, but also for civilization. Towns and cities, agriculture, livestock, industry, power, and many other prerequisites of modern life rely on access to dependable and ample water supplies. This has always been the case. From the very beginnings of organized social life, water has been an essential requirement. All of the first great civilizations were born on the banks of major rivers. Bounded by the Tigris and Euphrates Rivers, Mesopotamia or the "land between the rivers" in ancient Greek, gave rise to ancient Sumer, the first civilization in the region and possibly the entire world.[24] Mesopotamia was also subsequently home to the Akkadian, Babylonian, and Assyrian empires and it was these societies that created the first written language known as cuneiform, invented the wheel, and developed the first systems of irrigation. On the border of what is now India and Pakistan, the Indus River valley produced the Harappan civilization, whose largest cities boasted brick homes with bathrooms, indoor plumbing, sewer systems, reservoirs, and granaries.[25] At its peak, this civilization covered around three hundred thousand square miles.[26] Most famous of all, ancient Egypt owed its existence to the Nile, whose annual floods nourished a thin strip of land on either bank that allowed for the cultivation of crops and the development of a complex society whose archeological legacies are still considered marvels of engineering and a tribute to the power of the pharaohs.[27] Regular as clockwork and coinciding with the rhythms of agriculture, the annual flooding of the Nile covered the fields with a fertile black silt that allowed for rich harvests year after year.

All of these "hydraulic civilizations," as they have been termed, owed their existence to a reliable source of freshwater for crops, sustenance, power, and transportation.[28] Usually authoritarian in nature, the power of the ruling elite was always based on control of water, and when the water failed, as it was sometimes prone to do, so too did the civilizations built upon its supply. Since political, social, economic, and religious power rested on controlling access to water, when the rains didn't come or the river failed to flood, social unrest and societal collapse often followed. Evidence suggests, for example, that the Harappan civilization died, in large part, because of a long-term and severe drought that ended reliable monsoon rains and depleted local rivers.[29] Even fabled Egypt suffered when the monsoon rains failed to appear in the highlands of Ethiopia: the place where the headwaters of the Blue Nile and its tributaries are located. When the rains didn't come, the Nile didn't flood and during those periods archeological evidence suggests that many Egyptians starved while others became so desperate that they resorted to cannibalism. Water, in short, was crucial for the rise of civilization, and when once reliable sources of freshwater weakened or failed, those same civilizations also weakened and sometimes died. Water has always been the one indispensable resource upon which all life on earth depends. Water, as they say, is life, but not all water is the same and not all will serve our needs.

We humans require fresh or potable water. "Potable" simply refers to water that is pure enough to drink and use in food preparation. Water that is contaminated, polluted, salty, or brackish can cause sickness, disease, and death. Even though we live on a water world in which over 70 percent of the earth's surface is covered with the stuff, the simple fact is that we are running out of *usable* water. Ironically, most water is not suitable to sustain human life. Over 97 percent of the earth's supply of water is saline and therefore unusable for drinking. Less than 3 percent is freshwater. Of that miniscule 3 percent, most of it is largely inaccessible as snow, ice, groundwater, or moisture in the soil. In other words, the vast majority of the world's freshwater is locked away as ice in glaciers and at the poles or else is hidden in aquifers and pockets underground. Of all the water on earth, only about .3 percent is potable surface water. That's it—just .3 percent is all that is readily available. All of humanity largely depends on that miniscule proportion of the earth's water supply to sustain our civilizations, our communities, and our lives. Given its scarcity, we shouldn't be surprised that access to usable water is already a significant problem in many parts of the world, especially when we understand that 85 percent of the world's population lives on the driest half of the planet.[30]

According to the United Nations, almost 800 million people around the world do not have access to supplies of clean water, while another 2.5 billion do not have sufficient sanitation for sewage and waste disposal, a condition lending itself to the development and transmission of disease.[31] Sanitation is largely about keeping human waste, especially feces, away from human contact and the water supplies we use for bathing, cooking, and drinking. The World Health Organization suggests that almost two billion people use water sources contaminated by feces, which transmit diseases such as dysentery, cholera, typhus, and polio.[32] Such illnesses are estimated to kill around half a million people every year.[33] In sub-Saharan Africa, 62 percent of the population does not even have access to a simple toilet or lavatory and this deficiency often translates into contaminated water supplies. Some have even suggested that diarrhea kills 1 child every 18 seconds, 200 children an hour,

4,800 children every single day.[34] So extensive is this problem that it turns out that the lack of adequate or clean water is the leading cause of death in the entire world, especially among children, with much of that due to the death toll in developing nations.[35] Furthermore, the World Health Organization predicts that by 2025, "half of the world's population will be living in water-stressed areas."[36] This has significant implications for human civilization since, as James Salzman reminds us, "for more than three thousand years, understanding and management of drinking water have dictated the growth and health of human settlements."[37]

Unfortunately, the problem is only becoming more acute because of population growth, urbanization, and rising standards of living in many parts of the world, all of which contribute to increased demands for water. Whether it is nicer homes, more clothes, cars, and other conveniences of modern life, or better and more varied food or simply more of it, the reality is that the higher the standard of living people enjoy, the more they demand, and that requires water. We often don't realize that producing consumer goods and food requires a great deal of water during the refining, production, and manufacturing processes, not to mention the water used to irrigate crops. In fact, the principal culprits in our increasing demand for water are agriculture and industry. Agriculture alone takes up 70 percent of the available freshwater supplies, while industry requires another 20 percent.[38] Currently, the world population stands at about 7.3 billion human beings and continues to grow. Such a continual population expansion requires ever more water for consumption, crops, and the manufacture of goods. Between 1960 and the year 2000, the world's population doubled from three billion to six billion. During the same time period, annual grain production almost tripled, but it came at a cost in terms of heavy reliance on irrigation, which—when coupled with scientific advances in breeding and cultivation, the widespread use of agricultural machines, and heavy reliance on pesticides—allowed for such a dramatic increase in yield.[39] To feed everyone in 2012, the amount of land devoted to agriculture was over four billion hectares, which represents just under 40 percent of all the ice-free land on earth.[40] Yet according to the United Nations's Food and Agriculture Organization, it is water, not land, that limits the amount of agricultural production.[41]

Ironically, although it might seem counterintuitive, a warmer world with more carbon dioxide in the atmosphere does not necessarily translate into better harvests. Recent research suggests that while such conditions do cause many crops to grow faster, they also reduce grain yield for those same crops. This includes corn, rice, and wheat, the big three when it comes to feeding the world's population.[42] Having so much land under cultivation requires incredible amounts of water; irrigated farmland is much more productive than that which relies solely on rainfall. One important source of this water comes from aquifers. Aquifers are underground zones of permeable rock, gravel, or sand that are saturated with water. Accumulating slowly over time from rainfall and snowmelt seeping through the soil, these underground deposits of water have been around for thousands of years and recharge at glacially slow rates. Because of the relentless need for water in agriculture, these underground reservoirs are being depleted much faster than they are capable of being refilled through natural processes. India, for example, has drilled more than twenty million wells to access these supplies and now withdraws five times as much groundwater as it did in the 1960s.[43] In some parts of the country, the effects of this overuse have combined with drought and higher temperatures to severely impact agriculture and have created severe hardship for the small farmers who produce most of India's crops. Over 330 million Indians have been impacted by this crisis and more than 12,000 farmers killed themselves in 2014 because the situation was so dire.[44] Between 1997 and 2007, estimates suggest that around 150,000 Indian farmers committed suicide because of crop failure and debt.[45] Clearly, Indian farming is in serious crisis, but Indian society is not alone. The aquifer beneath Mexico City, for example, has been depleted to such an extent that parts of the city have subsided over seven meters resulting in many buildings being damaged or destroyed, and streets and structures tilting and undulating wildly as areas of the city subside unevenly.[46] Furthermore, around a fifth of the city's population does not have access to tap water and must instead pilfer or buy drinking water. In Iran, groundwater supplies have run out completely in some areas and resulted in entire communities needing to be abandoned.[47] The capital of Yemen, Sanaa, is facing a similar situation since evidence suggests that it is using up its groundwater supplies at four times the natural replenishment rate and is expected to completely run out of water by 2025 if current water use patterns continue.[48] These are just a few examples of the ways in which water shortages are cropping up with increasing frequency around the world. The era of easy water has clearly ended.

Ironically, climate change is projected to impact water in two primary ways that while appearing to be polar opposites are in actuality closely linked. Some locations will receive too much water in the form of flooding, while others will experience drought. Flooding will occur because of heavy rainfall from stronger storm systems and because of sea level rise brought about by the melting of the world's ice and through thermal expansion. Glaciers and the ice sheets of the Arctic, Antarctica, and Greenland are melting at unprecedented rates and a number of studies have recently come out indicating that sea levels have been rising faster than initially predicted and could rise more than four feet during the twenty-first century.[49] In just the past four years, over a trillion tons of ice has melted in Greenland alone and as one climate scientist warns:

> What concerns me the most is that this is the kind of experiment we can only do once. A lot of people don't realize that. If we start opening the floodgates on some of these glaciers, even if we stop our emissions, even if we go back to a better climate, the damage is going to be done. There's no red button to stop this.[50]

One study, in fact, suggests that the extent of sea level rise could be even higher because it suggests that the melting of the Antarctic, Arctic, and Greenland ice sheets has been underestimated in many studies.[51] Whereas in 2013 most scientists agreed that by the year 2100 sea levels would rise no more than three feet, many are now suggesting a rise of six to seven feet by the end of this century.[52] One recent estimate went so far as to suggest that sea level rise might happen much quicker than previously thought and could reach ten feet by 2065.[53]

Regardless of the exact amount of sea level rise, the impacts will be devastating. Low-lying coastlines will be completely inundated, while slightly higher land areas out of the immediate reach of rising sea levels will be subject to periodic flooding due to higher tides and storm surges. Storm surge refers to the actions of large storms such as hurricanes pushing large amounts of water in front of them and sending the water deep inland. In 2012, for example, Hurricane Sandy hit the eastern seaboard of the United States and sent a six-foot-high storm surge onshore into New Jersey and New York City, flooding entire coastal communities, destroying homes, and submerging tunnels and subway lines. The flooding caused electrical explosions and led to many fires. Almost five million people lost power and the New York Stock Exchange and all public transit systems were forced to shut down. The storm killed more than one hundred people and resulted in $50 billion to $65 billion in damages.[54] Keep in mind that more frequent and more extreme weather events, such as storms, are predicted as one consequence of climate change, which means that the kind of flooding brought about by Hurricane Sandy will in all likelihood become more common.

While the world's oceans have always been rising and falling depending on the amount of glaciation present in the world, what makes the present situation different is the speed at which sea levels are rising and that the world today has far more people—many of them living close to coastlines—spread across its surface. According to many estimates, approximately 40 percent of the world's population now lives within sixty miles of the coast.[55] Furthermore, coastal areas are experiencing much higher rates of population growth and urbanization than more inland regions,[56] with many of the world's largest cities located in coastal regions.[57] Particularly prone to flooding are coastal estuaries and river deltas. Deltas are simply low-lying deposits of sediment that fan out into the ocean from a river's mouth, and they have seen much population growth in recent years.[58] Of the thirty-two largest cities in the world, twenty-two are located in estuaries and deltas including London, New York, Shanghai, Buenos Aires, and Alexandria.[59] Around 300 million people inhabit these delta regions and these individuals will be among those most affected by the storm surges, higher tides, and flooding brought about by the increase in sea level.[60] If we look at China, for example, we find that the bulk of the population is situated in a crescent along China's eastern coastline, which, not coincidentally, is also where the vast majority of China's economic infrastructure is situated in the cities, ports, and factories that have helped power China's economic growth.[61] All of these will be at significant risk in a warmer world and will require a massive investment to mitigate the direct impact of sea level rise. To compound the difficulties, while many Chinese in the coastal regions have benefited from their country's economic development, most Chinese in the interior remain mired in poverty; such inequality has often resulted in tension and conflict that could reappear and strengthen as the country struggles with losses to its economic infrastructure, as well as massive potential internal population displacement.[62] China is home to more than fifty separate ethnic groups that include Han, Zhuang, Manchu, Hui, Miao, Uyghurs, Yi, Tujia, Mongols, Tibetans, Buyei, and Koreans, as well as a variety of religions including Confucianism, Buddhism, Christianity, Islam, Judaism, Bon, and Dongbaism. Such diversity of belief and identity has often provided the fault lines around which ethnic and religious violence has developed in response to crisis and it is quite possible that ethnic nationalism and religious intolerance will once again reappear and lead to violent conflict. China, however, is not alone in facing such risk from flooding. Many vulnerable cities around the world are waking to this danger and beginning to explore and invest in efforts to minimize the projected impacts. This process has accelerated and expanded recently as evidence continues to mount that this is not merely a hypothetical future concern but is already here.[63] Scientists in the United States have reported a dramatic increase in what is often referred to as "sunny day" flooding, leading one scientist with the National Oceanic and Atmospheric Administration to assert, "It's not a hundred years off—it's now."[64]

Unfortunately, there are no simple solutions since sea level rise and higher tides and surges will spread beyond the immediate impacts brought about by coastal inundation. Coastal freshwater supplies depended upon by millions of people for consumption and agriculture will be threatened as salt water increasingly infuses into the rivers, bays, and coastal aquifers. Salt also percolates through coastal soil in a process known as salt intrusion, rendering that land unable to grow crops. Many key agricultural areas in Asia lie along major river deltas and are highly susceptible to sea level changes, as are the people and communities that depend on those crops. Fishing communities will also be negatively impacted because the coastal fisheries they depend on for their livelihoods and survival will be lost due to warmer waters.[65] In countless direct and indirect ways, higher sea levels will profoundly impact the ability of coastal communities and nations to cope and deal with the challenges and changes demanded of them. But flooding due to sea level rise is not the only type of inundation that threatens. Flooding from rainfall will also pose a problem, and ironically, will often dispro-portionately affect drought-stricken regions.

While we generally think of rain as something positive, it can also have detrimental consequences. Too much rain can be as harmful as too little since harder and more frequent rains can result in rivers spilling over their banks and flooding communities and fields. Furthermore, in a warmer world, precipitation that once came down as snow will more often fall as rain and lead to potential inland flooding. California, for example, is predicted to be much drier on average in the coming years, but will also be prone to much more flooding because of this phenome-non. During the winter months, the snow that once created the snow-pack in the Sierra Nevada mountains will now descend as rain and inundate streams and floodplains. California has actually suffered from a number of megafloods in the past due to such conditions. During the winter of 1861 and 1862, for example, a massive amount of rain melted the snowpack in the mountains and, because the ground was frozen, was unable to be absorbed. Rivers swelled into massive torrents of water that flooded the entire Central Valley of California and created an inland sea that was about four hundred miles long and sixty miles wide.[66] Keep in mind that dry soil has a reduced capacity to accept and absorb water. In dry regions, rain very quickly overwhelms the ability of the ground to absorb the water, which just runs off. This is why flash-floods are common in desert regions when monsoon rains hit. In Arizo-na, where I live, people are often cautioned against camping in dry river washes even though they appear to be little more than sandy gullies. They are dangerous during the monsoon season because sudden show-ers can send a wall of water racing down a wash, sweeping away every-thing in its path within minutes, even if the rainfall was miles away.

As with other aspects of climate change, when it comes to rain there will be winners and losers. Large parts of the American West, and vast swaths of Africa, Asia, and Australia, will most likely see droughts inten-sify and lengthen in duration. Precipitation in these areas, whether rain or snow, will tend to be less frequent and more sporadic with important consequences for agriculture, industry, and communities in the affected areas. Monsoonal rainfall patterns will also vary, in some cases becom-ing less regular and less ample, while other regions will see a boost to their average annual rainfall. India, for example, is expected to be par-ticularly hard hit since more than 60 percent of its population depends on monsoon rains for subsistence-level farming and those monsoonal rains are expected to weaken substantially and become much less regular, raising the specter of widespread famine reappearing in the subcontinent.[67] Furthermore,

> the melting of the snow from the Himalayan glaciers means that India's major rivers—especially the Ganges, its tributaries, and the Brahmaputra—could alternate between abnormally low flows in the early summer and winter months and extraordinarily high flows during the monsoon, posing the double risk of drought followed by flood.[68]

It's easy to see how such a situation could contribute to a destabilization of Indian society as widespread famine, water scarcity, and disease confront ever larger segments of the population. Two-thirds of Indians are farmers who will see their livelihoods and ability to feed themselves and their families severely compromised. In the face of widespread famine, disease, and loss, will we see, as Christian Parenti suggests, "a catastrophic convergence of climate, poverty, and violence?"[69] With a long history of religious, political, and ethnic tensions between Hindus, Muslims, and Christians, India is already dealing with homegrown insurgencies and rebel movements that could threaten to destabilize Indian society even more.[70] Given the frequent clashes in the past, it is not unlikely that such historic antagonisms and conflicts may be resurrected by desperate communities and unscrupulous political, social, and religious leaders eager to exploit the situation for their own ends or simply to act on their own prejudices. Such a scenario is already a concern in the wake of recent outbreaks of communal violence. On September 11, 2016, for example, an Indian court ruled that the Indian state of Karnataka would have to release water to the neighboring state of Tamil Nadu, and in response, violent protests and riots broke out in Karnataka over this perceived injustice, much of which played out along ethnic identity lines.[71] Water in these Indian states was already in high demand since local rice fields in those states require a great deal of it and also because it had been a weak monsoon year. The court ruling, coming as it did at an inauspicious time, fueled violent reactions in a setting where access to water is often literally a life and death issue. Clearly, the connection between water and conflict is not merely an exercise in hypotheticals.

Historically speaking, water has often been an underlying source of conflict in many places around the world, a point Michael Klare makes when he reminds us that "governments have repeatedly gone to war over what they view as 'vital national interests,' and these include oil and water supplies."[72] In 1967, for instance, Israel launched a preemptive strike against Egypt, Syria, and Jordan, largely in response to a series of escalations and provocations, especially the mobilization of Egyptian troops along Israel's border in the Sinai Peninsula.[73] Ariel Sharon, Israel's commander during that conflict, later wrote that the war actually began two and a half years earlier when Syria began construction of a canal to divert water from the tributaries of the Jordan River: a source of water vital to Israel. Sharon specifically suggested that "while the border disputes were of great significance, the matter of water diversion was a stark issue of life and death."[74] In a largely arid land, control of water has always been crucial to Israel's survival and is a large part of the reason Israel has not been willing to return the West Bank or the Golan Heights after occupying those territories. Those regions control access to a key aquifer for the entire region and are vital to Israel's survival.[75] The Indus River and its tributaries provide another example of water-based conflict.

Beginning in the high Himalayas, the Indus flows through the northwestern corner of India and from there into Pakistan where it supplies 90 percent of the water used for crops.[76] The Indus is literally the lifeblood of Pakistan, one of the most water-stressed countries in the world.[77] The water situation in Pakistan is worsening with wells running dry and groundwater levels declining dramatically, by about fifteen to twenty feet, in many places.[78] Per capita, access to water has declined by 75 percent, due largely to overuse because of rapid population growth and every year one hundred thousand acres of farmland are abandoned because of salinization as a result of irrigation.[79] Consequently, many thousands, if not millions, of Pakistanis have left their rural fields and communities in order to take their chances in the burgeoning cities while others have resorted to banditry and extortion. During the 1990s, for example, the city of Karachi was the fastest-growing city on earth and is now home to over ten million residents.[80] Importantly, many Pakistanis hold India responsible for their water predicament since the river first flows through the Indian states of Jammu and Kashmir and in order to meet its own needs the Indian government has constructed a number of dams on the Indus and its tributaries to divert water toward Indian farms.[81] In some years, the Indus doesn't even reach the sea, and these Indian water diversion initiatives have served to exacerbate tensions dramatically in the region. One extremist group in Pakistan, Lashkar-e-Taiba, has gone so far as to label Indian dam building and water diversion projects "water terrorism."[82] It doesn't help that there is a long history of violence between these two nuclear powers; they have fought three major wars and engaged in many smaller skirmishes over a variety of issues. Unfortunately, it does not appear as if Pakistan has the capacity to easily cope with the challenges posed by water scarcity. While its population continues to grow (expected to top 310 million by 2050, making it the sixth-largest nation in the world by population),[83] its institutions and infrastructure have not developed at the same rate. One commentator has noted: "At present Pakistan's state institutions are relatively weak and unlikely to serve as bulwarks against the challenges of climate change. The country's feudal and tribal society social structures are not typically viewed as reserves of social resilience or ingenuity, but rather of repression and limited human capital."[84]

Pakistan, in other words, has low resiliency. But the link between water and conflict is not always about fights over access to supplies of water. The connection can also be more indirect.

Drought, for example, can negatively impact the agricultural output of a region and result in food shortages that trigger protests, rioting, and violence. In a three-year period, between 2005 and 2008, the price of wheat, corn, and rice shot up on the international market; wheat and corn suddenly cost three times what they had previously, while rice saw a fivefold increase in cost.[85] These three staples provide the main source of food for the world's population and this drastic increase in prices hit poor people particularly hard. So central is bread to Egyptians, for example, that their word for it—*aish*—is also the word for life. In the aftermath of these global cost hikes, protests and riots broke out in Egypt, Haiti, Cameroon, Bangladesh, Bolivia, Burkina Faso, Côte d'Ivoire, Mexico, Yemen, Pakistan, Uzbekistan, Sri Lanka, Senegal, and Somalia.[86] Other countries, such as Russia, Vietnam, India, and Thailand, took strong steps to prevent such unrest and began freezing prices or banning rice exports.[87] While the price increases and the violent protests were relatively short lived in these instances, one can envision a scenario where the situation does not improve and food scarcity triggers much more sustained and widespread violent reactions from those for whom such issues are truly a matter of life and death.

One recent study on global warming and civil war in Africa found that climate change increases the risk of civil war with the primary mechanism connecting the two being a decline in agricultural production.[88] Keep in mind that while agriculture only accounts for 2 percent of the GDP in the developed world, it composes 11 percent in the developing nations, and fully 40 percent in Africa.[89] Into the mix we can also throw the issue of population displacement—a subject to which we will turn in the next chapter—as those negatively impacted move away from the stricken region in search of hope and economic opportunity, but in the process triggering further social unrest or increasing tension and conflict in the host nations from those who resent or fear the influx of incomers. One expert on water issues puts it this way:

> As in history, future human migration patterns will be influenced by water availability, or the lack of it. A large exodus of people from water-scarce regions to water-sufficient areas will swamp the latter socially and economically, potentially triggering local backlashes and straining their internal security and environmental sustainability.[90]

The impact of climate change on water and its connection to violent struggle and conflict is not simply an abstract theoretical exercise about future possible scenarios. It is something that is present here and now. Some have argued that the genocide in Rwanda, for instance, had environmental undertones that helped shape and facilitate its perpetration.[91] While the proximate causes of the genocide had a lot to do with the legacy of colonialism; a history of prejudice, persecution, and violence between the Tutsi and Hutu population groups; ethnic nationalism and politics; and the civil war being fought in the early 1990s between the Hutu-led government and the largely Tutsi-led Rwandan Patriotic Front, other less evident factors also played a role.[92] Essentially, Rwanda had been experiencing recurrent droughts that had served to reduce the productivity of crops. This was a serious problem given that Rwanda was a densely inhabited society with arable land in short supply. Most of the rural inhabitants were subsistence-level farmers and in the radicalizing crisis of the civil war, amid heightened ethnic antagonisms, the genocidal impulse was more easily imagined and unleashed by political leaders and more willingly supported by ordinary Rwandan Hutu farmers. The case of Syria also underscores the reality of this linkage between violence and water, but these are not the only examples. Genocide and water scarcity are also at the heart of the world's first genocide of the twenty-first century in the Darfur region of the Sudan, and to understand that situation we must first understand the Sudan's relationship with the Sahara.

The Sahara is the world's largest nonpolar desert and covers almost the entire northern portion of Africa. The Sahara is so big that the United States would fit easily within its borders. Covering almost four thousand square miles, it stretches from the Atlantic Ocean in the west, to the Mediterranean Sea in the north, and to the Red Sea in the east. To the south is a semiarid savannah that leads into the Sudan and the rest of sub-Saharan Africa. When we think of the Sahara we tend to think of endless rows of towering sand dunes and relentless heat—in fact, the hottest temperature ever recorded on earth was in the Sahara with an astonishing 136 degrees Fahrenheit; yet it was not always so. Once this region was much more temperate, watered, and even lush. Prehistoric petroglyphs and pictographs show elephants, rhinos, and other large game in Saharan locations that are now windblown wastes but which were once able to support such animals. Arrowheads and flint knives have been found in desolate Saharan regions where hunters once roamed in search of prey.[93] Archeologists have even found fishhooks in regions where only sand dunes now exist, and photos and radar images taken from space show the faint outlines of past rivers and lakes across many parts of the Sahara.[94] Brian Fagan describes the Sahara as a vast ecological pump, which is a poetic way of suggesting that during wetter eras, the Sahara retreated and allowed people, plants, and animals to push into the interior, while drier times resulted in the sands advancing and pushing them back out again. This is an age-old cycle that has played out over countless eons and continues to shape the modern interplay of environment and human community. This connection holds true for the genocide that began in 2003 in Sudan.

About the size of Europe, Sudan or Land of the Blacks, as it was dubbed by Arab geographers in the Middle Ages,[95] is one of the largest countries in Africa and sits on the northeast shoulder of the continent. Bordered in the north by Libya and Egypt, Chad and the Central African Republic in the west, South Sudan in the south, and Ethiopia,

Eritrea, and the Red Sea in the east, large parts of the Sudan are desert, especially in the northern and central parts of the country. In Sudan, the farther south you travel, the more rainfall you get, and the greener and more lush the landscape. Running like a lifeline down the middle of the country is the Nile. When entering the Sudan, the Nile is actually two separate rivers with the headwaters of the White Nile far to the south in Lake Victoria, and the source of the Blue Nile to the southeast in the highlands of Ethiopia. Both rivers meet in the heart of the Sudan and from there the newly combined river flows north toward Egypt and its eventual rendezvous with the Mediterranean Sea. At the confluence of the two rivers lies the ancient capital of Khartoum, long the seat of the nation's political, religious, and economic power. The modern state of Sudan was formed in 1821 on the heels of an invasion by an Egyptian army that overthrew the existing sultanates and brought the region under the control of the Ottoman empire,[96] and it was during this era that the pattern was first established of developing the Nile River basin to the neglect of other portions of the country, a disregard that was to bear bitter fruit in later years. During the 1880s, an Islamic revolution led by Mohamed Ahmed, better known as the Mahdi or Expected One,[97] established a strict fundamentalist version of Islam that "sought to cleanse the country of foreign influences" and that "would make periodic reappearances through Sudanese history, with violent consequences for the Sudanese people."[98]

A significant part of the problem was that historically the tribes of the Sudan were perceived as being either African or Arab, even though there are no real racial differences between the tribes.[99] Those tribes from the Nile River valley, where much of the political power and economic development were centered, tended to be much more Arabic in language, culture, and religion than those living in other parts of the Sudan, although an often heterogeneous mix of tribes could be found in many regions of this vast country. Importantly, those defined as Arabic have often looked down on those tribes defined as African and treated them as inferior and less civilized even though they are fellow Muslims.[100] In part, this antipathy stemmed from the fact that many tribes in the Darfur region have retained much of their traditional cultures, languages, and beliefs, despite having converted to Islam.[101] In many ways, the simplistic dichotomy of Arab versus African conceals a more complicated mix of religion, culture, lifestyle, and traditions that have often shaped and influenced Sudanese politics, economics, social life, conflict, and violence. In short, a concentration of wealth and power in Khartoum and adjacent areas to the neglect of regions on the periphery and the establishment of a strict puritanical version of Islam combined with deep tribal-based schisms and racism created a toxic national environment that was to have profound and deadly consequences at the turn of the century in the Darfur.

The Darfur or Land of the Fur—the Fur being a large tribal group living in the region—refers to the western part of the Sudan and is a semiarid, high-altitude plain that covers around 150,000 square miles and since 2003 has been the scene of the twenty-first century's first genocide. Covering an area as large as France, this region is roughly divided into North Darfur and South Darfur. North Darfur, abutting as it does the Sahara, is an arid and inhospitable region that is home to pastoralist tribes whose seminomadic lifestyles revolve around their livestock. They also tend to be perceived and treated as Arabic rather than African. Given the dry conditions, these peoples and their herds have often been vulnerable to climatic variations. During drier times, the Sahara moves south and encroaches on already fragile grazing lands, before receding once again in wetter periods. South Darfur, on the other hand, tends to be more hospitable, which is why during times of drought, there have often been large-scale migrations of people from North Darfur to South Darfur.[102] The majority of the tribes in the south, such as the Fur, Masalit, and Zaghawa, are commonly defined and treated as African and generally practice agricultural lifestyles with fields, small farms, and towns scattered throughout the region. Over time, these periodic dislocations of Northern Darfurian tribes southward resulted in South Darfur achieving a population twice the size of North Darfur with a consequent strain on land and water resources.

The spark that set the current round of violence in motion was the publication and distribution of thousands of copies of the first half of *The Black Book: Imbalance of Power and Wealth in Sudan* in May 2000, with the second part following in August 2003.[103] The intention of the anonymous authors was laid out explicitly when they wrote, "This publication unveils the level of injustice practiced by successive governments, secular and theocratic, democratic or autocratic, since the independence of the country in 1956 to this date."[104] Essentially, the *Black Book* was an indictment of the hard-line National Islamic Front government that had taken power in a coup eleven years earlier. In meticulous detail, this manuscript pointed out that most of the economic and political power and opportunity in the country were concentrated in the hands of a small number of people from three specific tribes that came from the area surrounding the capital city of Khartoum. According to the statistics compiled in the book, almost all government officials, civil servants, high-ranking military and police officers, judges, and bank officers, for example, were drawn from members of those three Arabic tribes.[105] Furthermore, this publication also pointed out that the vast majority of doctors, most of the nation's industry, and almost all investment were similarly based in the capital of Khartoum. Darfurians, in other words, and most other Sudanese, had been overwhelmingly marginalized and excluded for many years from the centers of power, economic opportunity, and access to vital services such as medical care. Throughout the *Black Book*, there appeared a constant demand for "justice and equality,"[106] and partially in response to this call for action, members of several tribes in the Darfur region formed a number of rebel groups—the Sudan Liberation Army and Movement (SLA/M) and the Justice and Equality Movement (JEM)—and began an insurrection by attacking a variety of government installations, in particular, a number of police stations and military bases.

The extremist government in Khartoum reacted with stunning brutality and viciousness by organizing a genocidal campaign led largely by government-supported paramilitary groups. They began by actively recruiting young men from Arab tribes, especially those from North and South Darfur, and when volunteers showed up from non-Arab tribes, they were turned away.[107] These militias, known collectively as the Janjaweed, began attacking and destroying villages and communities throughout the Darfur. The pattern was usually the same. First a government plane would appear over a village and bomb the community. Since the government didn't have any real bombers, they relied on transport planes and simply rolled off oil drums filled with explosives and metal debris to act as shrapnel.[108] As Gérard Prunier points out, such bombs are largely useless against military targets, but extremely effective against stationary and undefended civilian communities.[109] After this initial bombing, government attack helicopters would arrive and machine-gun schools and any other large buildings left standing. Generally, it was at this point that the militia groups would roll in to complete the destruction. Sometimes they were accompanied by regular army troops while other times they were on their own. Typically, they would be mounted on horses and camels, or riding in the back of modified pickup trucks known as "technicals." These paramilitary forces would surround the village, and then loot, rape, and murder those who hadn't already fled or been killed. At the end of their violent assault, they would conclude by setting fire to the remaining houses. Other times, instead of killing most of the villagers, they would only murder a few of the men, rape many of the women, and then brand their victims in order to disfigure them. Such horrific scenes occurred in countless small communities and towns throughout South Darfur. In trying to explain such a relentless and genocidal campaign initiated by the government in Khartoum, many have focused on the role of the Arab-versus-African tribal divisions in Sudanese society. The extremist government in Khartoum had long marginalized Darfurian tribes because of the perception that they were more African than Arab and therefore inferior, and the rebel attacks in 2003 gave the government free rein to act on these prejudices.[110] One survivor of such an attack illustrated this by recounting:

> The Janjaweed were accompanied by soldiers. They attacked the people, saying: "You are opponents to the regime, we must crush you. As you are Black, you are like slaves. Then the entire Darfur region will be in the hands of the Arabs. The government is on our side. The government plane is on our side, it gives us food and ammunition."[111]

Such sentiments certainly played a role in generating the genocidal violence. One could also point out that the government had been fighting an on-again/off-again civil war against separatists in the south of Sudan, and that the long-standing conflict had only just ended with a cease-fire when the insurgency in Darfur broke out. Given the timing, the government acted quickly and ruthlessly to stamp out the rebellion in Darfur so that it would not result in a protracted war similar to the one waged against the tribes in the south. Furthermore, one could also suggest that the long-standing war in South Sudan had radicalized the government into more easily accepting violent and extreme solutions to problems. While all of these arguments have merit in explaining why genocide erupted in Darfur in 2003, we shouldn't ignore the underlying geographic and climate-related factors that played a crucial role in creating a context in which genocidal ideologies and policies could develop and be implemented.

Specifically, the Darfur region of the Sudan had long been suffering the effects of overgrazing and drought.[112] This was, as noted above, an issue that periodically cropped up and resulted in many northern Arab tribes moving from North to South Darfur. Remember that many of those from the northern part of the Darfur were culturally more Arabic and lived seminomadic lives based on their livestock, whereas the more African tribes living in the south were much more agricultural and sedentary in their lifestyles.[113] During the 1980s, the influx of nomadic tribes from the North had resulted in more and more conflict with their South Darfurian neighbors as a drought worsened the problem of overgrazing and created famine conditions in many parts of North Darfur. This forced many nomadic northerners to migrate into the more agricultural southern part of Darfur and eventually led into open conflict between the various tribes.[114] Interestingly, it was during this era that the Janjaweed tribal militias first made their appearance,[115] but it wasn't the first time such environmental conditions had created intergroup conflict. Drought and famine in the 1960s and again in the 1970s had also resulted in intertribal conflict and violence.[116] Andrew Natsios summarized the situation well when he noted: "Historically, when famines strike anywhere in the world, they sometimes bring with them ethnic conflict, violent crime, and population movements of starving people seeking food, often leading to political upheaval. All of these struck Darfur."[117] Negotiations ended these first bouts of fighting but did not end the competition for increasingly scarce water and land, nor did it end the antagonisms between the tribes. One scholar tallied over forty conflicts between various Sudanese tribes over the years and highlighted the fact that water and grazing rights were the single most common basis of the violence.[118] When fighting broke out in 2003, the government in Khartoum was able to exploit this situation by recruiting the Janjaweed militia from seminomadic Arabic tribes in the northern part of the Darfur who were willing and able to participate in the genocide because it allowed them the opportunity to acquire land and water for their livestock by killing and displacing the tribal groups that stood in their way.[119] While political and military calculations, ethnic stereotypes, and long-standing antagonisms were certainly key factors in creating this genocide, we cannot ignore the role that drought, desertification, and overgrazing played in facilitating a situation in which the genocidal impulse could take hold. In this situation, resources, specifically land and water, were at the heart of the violence.

As of the writing of these lines in 2016, the genocide in Darfur continues, albeit in modified form. It was Eric Reeves, the well-known chronicler of the genocide in the Darfur, who as early as 2005 pointed out that the overt violence of that genocide had dwindled and was being replaced by something else. He wrote, "Genocidal destruction became more a matter of engineered disease and malnutrition than violent killing. . . . There came a point . . . in which the ongoing genocide was no longer primarily a result of direct slaughter, but of a cruel attrition."[120] This genocide by attrition, as it is now often referred to, continues to take a deadly toll on life in the Sudan and provides a powerful and tragic lesson as to ways in which climate change and water-related issues can lead to communal violence, war, and genocide. One important by-product of the genocide in the Darfur has been the creation of large numbers of both internally displaced populations and refugees. Such population dislocation is an extremely common consequence of war and genocide and it is to the issue of forced displacement that we turn next.

5

FORCED DISPLACEMENT AND BORDERS IN A WARMING WORLD

New wars will be environmentally driven and cause people to flee from the violence, and, since they will have to settle somewhere, further sources of violence will arise—in the very countries where no one knows what to do with them, or on the borders of countries they want to enter but which have no wish at all to receive them.

—Harald Welzer[1]

In the 21st century the world could see substantial numbers of climate migrants—people displaced by either the slow or sudden onset of the effects of climate change.

—Michael Werz and Laura Conley[2]

It is not difficult to imagine that conflicts arising from forced migrations and economic collapse might make the planet ungovernable, threatening the fabric of civilization.

—James Hansen[3]

The wall between Mexico and the United States in Nogales, Arizona, is an imposing structure, although calling it a wall isn't technically accurate. Rather, it is a fence composed of rust-colored steel posts set about four inches apart that are anywhere from fifteen to eighteen feet high depending on the terrain. On the north or U.S. side of this barrier, dirt and paved roads typically run parallel to the fence in either direction and you can often see white-and-green Border Patrol SUVs cruising along the fence line or parked in silent watch. Mobile and permanent light fixtures abound and are situated in such a way as to flood the wall with bright light at night. In some places, houses line the side of the street directly opposite the wall and it makes you wonder how those homeowners view the erection of such an imposing barrier that now dominates their neighborhoods. It is a jarring sight to stand there in the bright Arizona sun and look at this barrier. Scanning east and west, you can see the rise and fall of the fence as it contours along the steep hills that make up the landscape in Nogales. It is intimidating in its angular and relentless solidity and is not something you really expect to see along this heavily traveled border. Mexico is a nation that countless thousands of Americans visit for vacation and business every single day and vice versa. As one scholar noted, "The U.S.-Mexico border is in many ways distinctive: it is the most heavily traveled land crossing in the world and also one of the most heavily fortified."[4] Why is this? The United States, after all, has long championed the ideals embodied in the Emma Lazarus poem inscribed on the Statue of Liberty that reads, in part:

> Give me your tired, your poor,
> Your huddled masses yearning to breathe free,
> The wretched refuse of your teeming shore.
> Send these, the homeless, tempest-tost to me,
> I lift my lamp beside the golden door![5]

Standing in Nogales, such sentiments seem more ironic than idealistic and certainly don't seem appropriate for the situation on the U.S.-Mexico border today. This wall seems more in keeping with something you would see at a border between two nations at ideological or military odds, such as the demilitarized zone (DMZ) that divides North and South Korea, or perhaps the borders that divided Soviet Bloc nations from Western Europe during the height of the Cold War. I remember visiting the divided city of Berlin as a kid and traveling by train through East Germany. To do so we needed to transit through heavily patrolled and militarized borders complete with barbed-wire fences, towers with spotlights, and armed guards equipped with dogs and automatic rifles inspecting the train. Crossing into East Berlin from West Berlin involved a similar transition as we passed from the western into the eastern zone. These sorts of childhood memories came to mind as I stood on the U.S.-Mexico border looking at this steel barrier that now separates the United States from a country that, until fairly recently, had enjoyed a much closer, warmer, and more open relationship but which now looms large in the American imagination as a source of violence and criminality.

According to the historian Karl Jacoby, the U.S.-Mexico border "has served as a screen onto which Americans have projected some of their deepest anxieties,"[6] and the wall serves as the most visible representation of those fears. This chapter is primarily focused on exploring how those fears and anxieties may escalate into conflict and violence against refugee populations as the numbers of people who've been displaced by climate change are expected to dramatically accelerate in the coming years. As we will discuss, these vulnerable populations run the risk of being scapegoated and persecuted as host nations struggle to come to terms with the rapid and large influx of newcomers and it is in the border zones that such tensions and conflicts will be most visible. In many parts of the world, including the United States, we are already experiencing a hostile backlash against those seeking to leave their homes in search of safety and opportunity.

Over the last fifteen years or so, the United States has constructed a series of fences and walls that span about 650 miles of the border and which were built at a cost of billions of dollars.[7] Given that the entire border is close to two thousand miles long, the fence is really just a series of walls and fence structures in high traffic areas, usually towns and cities. In some areas, the fence is composed of steel posts sunk into concrete footings, while in others it is made out of corrugated sheet metal. Some sections are built with the top angled outward, while elsewhere it is topped with barbed wire. These physical obstructions are supported by a large array of technology that now surveils those sections of the border in which no actual wall yet exists. These include electronic sensors, cameras, a fleet of drones, and ground penetrating radar.[8] In addition, the Border Patrol, an agency that has grown dramatically in recent years, also monitors and patrols the border enforcement zone, and their work is sometimes informally augmented by various vigilante groups that have been notable for their radical rhetoric and sometimes violent tactics.[9] The proliferation of fences in the more populated zones of the border has served to push the flow of migrants into the deserts that do not have the same physical barriers. In some desert areas, all that demarcates the border is a single strand of barbed-wire fencing, and while this might seem a significant oversight, we have to understand that the desert terrain and environment serve much the same purpose as the physical wall, but with often deadlier consequences for those attempting the journey.

The southwestern United States and the northern region of Mexico are located on a harsh but incredibly beautiful land. It is a rugged desert landscape that encompasses the Mojave, Sonoran, and Chihuahuan Deserts, each different and unique as only deserts can be. From the Joshua trees, creosote bushes, and sage of the Mojave Desert, the saguaro and organ-pipe cacti and jumping cholla of the Sonoran, to the yucca and agave bushes of the Chihuahuan, this region stretches from California in the west, through Arizona and New Mexico, and into Texas in the east. It is an unforgiving land as many who've tried to cross have tragically discovered. The stretch of desert between Tucson and Yuma is so treacherous and dangerous that early Spanish travelers named the system of trails heading north the "Devil's Highway" (El Camino del Diablo), or alternatively, the "Road of the Dead" (Camino del Muerto). From ancient times to the present, this passage has earned its reputation in the blood and suffering of those who dared to test its hazards. Nowadays, those travelers tend to be from Mexico and points farther south who are lured north by the prospect of jobs and a better life in the United States. Sometimes their journey ends with arrest, detention, and deportation, but all too often it terminates in the harsh wastes of the desert with a lonely and painful death from heat, dehydration, exposure, and exhaustion.[10]

In many ways, the militarization of the U.S.-Mexico border reflects the increasing hostility experienced by Latino groups in the United States. Although illegal immigration is certainly not new, recent years have seen a sharp backlash against new arrivals from Mexico and other Latin countries. Beginning in the 1990s, the United States experienced a surge in illegal immigration that continued into the early 2000s. Interestingly enough, climate seems to have been an underlying driver for many of those attempting the crossing. One study, for example, found that changes in crop yields in Mexico due to drought and other climate-related effects were positively correlated with migration into the United States.[11] During times of drought and other climate events, many farmers found themselves unable to sustain themselves and their families because of reduced crop yields, so the choice for many was simple: stay and starve or hazard the dangerous trip north to find better economic opportunities. In noting a similar displacement of the Tarahumara people, an indigenous population living within Mexico, one commentator suggested that "the Tarahumara had moved out of the area now. It was climate change as much as anything. There had been twelve drought years in the last fifteen. And it was becoming impossible for subsistence farmers to keep themselves alive."[12] Similarly, recent years have seen increases in migration from Guatemala, El Salvador, and Honduras that are also largely driven by climate change in the form of drought.[13] When thinking about the topic of immigration from Mexico and points south into the United States, climate is rarely considered as a factor influencing the movement of people north, yet that often is an important underlying consideration helping drive such migration. Beginning in 2008, the rate of attempted crossings dropped sharply, but unfortunately, the fear and rhetoric around the issue did not similarly decrease.[14]

In the United States, concerns about the influx of undocumented immigrants were fairly muted until the al-Qaeda attacks on the United States on September 11, 2001. After that deadly day, fears of terrorists entering the country through a border often perceived to be porous added a new urgency to concerns about the number of undocumented migrants coming into the country through the southern border.[15] Fuel to the fire was added by the horrific violence endured by Mexico in its war against the Narcos that began in December 2006, when the president of Mexico sent in the army to attack the drug cartels and in doing so unleashed a drug war that has shaken Mexico to its core.[16] Extensive coverage of the violence by news media ensured that Americans were exposed to countless stories detailing the abductions, assassinations, mass murders, beheadings, and similar outrages that characterized much of the violence. In 2008, the Great Recession further accelerated an attitudinal shift against immigrants, both documented and undocumented, due in part to the economic insecurity experienced by many in the wake of the economic collapse that was amplified by a deep unease on the part of many who feared and resented the changing demographic composition of the nation. Tougher economic times typically increase harmful attitudes, hate crimes, and other forms of violence targeting minority populations. In the United States, for example, negative imagery and stereotypes of immigrant populations have often increased during economic downturns, along with various kinds of violent victim-ization. This correlation between economic hardship and xenophobia has been observed at other times in the United States and in other countries around the world.[17] Simply put, the U.S. border with Mexico has become a potent symbol and focal point for the polarizing issues of refugees, migration, and population displacement,[18] but it's not just the United States that is confronting this issue, nor is it the only nation experiencing a political and social backlash against such newcomers.

In February 2016, I found myself in Berlin, Germany. Berlin in February is a bitterly cold and gray place, although in many ways the weather seemed quite appropriate for what I was doing. I was there conducting research and visiting some sites of historic interest related to the work I do on the Holocaust and genocide, and because of that it somehow seemed very appropriate to be experiencing this city in the cold dreariness of those short winter days. Unexpectedly, however, I found that immigration was consuming Germany's attention and nu-merous talk shows and news programs were covering the issue in some depth. On Berlin's streets, I saw at least one demonstration and a lot of graffiti about this topic. As I learned, this situation largely began in 2015 when Europe experienced a dramatic increase in the number of mi-grants and refugees seeking to reach a European Union (EU) nation. While many were simply hoping to escape a lifetime of poverty and forge a better life, many more had been forcibly displaced by conflict, war, and genocidal violence. The first warning sign for most Europeans was the gruesome discovery of seventy-one dead bodies in a truck that had been abandoned by the side of a highway in Austria in August 2015.[19] Most were Syrian refugees who had likely suffocated after being packed like cattle into the back of the tractor trailer truck and these particular deaths sounded the alarm that Europe was in the middle of a profound refugee crisis that had been going on for some time.

Seeking to escape the civil war and the widespread atrocities that had engulfed their country, the largest contingent of refugees were coming from Syria. The second- and third-largest groups were from Afghanistan and Iraq, respectively, both nations also embroiled in ongo-ing conflicts that had sent many fleeing from the violence.[20] Others came from places such as Eritrea, Nigeria, Somalia, Iran, and Bangla-desh. While many sought refuge in neighboring countries, such as Tur-key for the Syrians, many more aspired to attain safety and opportunity offered by a more distant Europe and to achieve this they were willing to risk everything. Most chose to attempt a crossing of the Mediterranean or Aegean Sea from Turkey, Libya, and Egypt in order to reach Greece or Italy. The European Union had previously settled on a deterrence policy to discourage migrants from attempting to enter its borders and had hardened some of the easier avenues of access into Europe, thus forcing potential refugees into the more dangerous sea passages.[21]

These desperate people negotiated with unscrupulous smugglers and traffickers who were just as likely to rob them as arrange passage and gambled their lives on often unsafe, inadequate, and overcrowded boats. According to the International Organization for Migration (IOM), almost four thousand died trying to cross the Mediterranean in 2015, but the flood of people kept coming.[22] One photograph in particular captured the horror and the tragedy of this modern-day exodus. In September 2015, an inflatable raft capsized on the passage from Turkey to Greece and one of the victims was a three-year-old Syrian Kurd child named Alan Kurdi. The picture of his young lifeless body lying facedown on a beach reverberated around the world and served as a stark and terrible reminder of the very real human cost this migration is taking on those who risk their lives on this perilous journey. Frontex, the European Border and Coast Guard Agency, estimates that about one out of every four who attempt to reach Europe by boat dies in the attempt,[23] thus making the European border the most dangerous one in the world. These numbers don't even capture those who have died trying to cross the Sahara to reach Europe, or who have died trying to cross the mountains in eastern Europe, or who have been killed in minefields on the Greek-Turkish border.[24] Amazingly, given the dangers, some have risked the dangerous voyage repeatedly, an eloquent testimonial to the perils in their home countries and the powerful allure of Europe. Wolfgang Bauer, a German journalist who attempted the dangerous journey undercover, recounts how he met a young man who was trying to reach Europe for the sixth time. This intrepid individual had experienced having his boat sink underneath him and needing to swim to shore in order to save himself; being repeatedly robbed by smugglers; and being arrested by the Greek Coast Guard and sent back to Turkey on two other occasions.[25] Yet when Bauer met him, he was attempting the crossing once again, so important was the prospect of reaching a European Union country.

For most newcomers, the first port of call was usually Italy or Greece. From there, most would try to make their way to the wealthier nations farther north and west. Because of the EU's Dublin Regulation stipulating that refugees must seek asylum in the first EU nation in which they arrive, these frontline nations were in danger of being overwhelmed by the logistics and costs of caring for so many asylum seekers. Violence sometimes accompanied this mass migration as patrol and coast guard boats occasionally fired on or rammed vessels with refugees aboard, sometimes with significant loss of life, while on the Hungarian border, asylum seekers were frequently attacked and beaten by border guards as they attempted to climb over newly erected barbed-wire fences. Belatedly, political leaders began to awake to the magnitude of the humanitarian crisis, and consequently, a number of EU nations geographically farther removed, agreed to waive the Dublin Regulation and accept many of these refugees. One of those was Germany, which gave entry to more than a million displaced people in 2015.[26] Such a large number of newcomers posed some serious issues for German society as towns and cities throughout Germany sought to absorb and cope with the massive inflow into their communities. Providing housing, employment, schooling, and medical care and processing applications for asylum imposed a tremendous strain on Germany's infrastructure and social welfare institutions, especially since most of the responsibility was left to local communities to handle.[27] Language, religious, and cultural differences served to compound the difficulties. Already struggling to meet the needs of this influx, the situation took a dramatic turn for the worse on New Year's Eve in 2015 when, in a number of cities around Germany, women were sexually assaulted by groups of young men, up to two thousand assailants by one count, many of whom were believed to be recent arrivals. One police report suggested that more than 1,200 women were victimized in the assaults that included unwanted sexual contact, groping, robbery, and rape. As Holger Münch, president of the Federal Crime Police Office, summarized in an interview, "There is a connection between the emergence of this phenomenon and the rapid migration in 2015."[28]

During my visit to Berlin in February 2016, I found the airwaves full of passionate and often heated debates and discussions about Germany's willingness to embrace so many refugees during the previous year. Germany is far more diverse today than it was when I was growing up there in the 1960s and 1970s, yet this increased diversity has not been embraced by all Germans. Strong and enduring undercurrents of suspicion and intolerance remain. One recent survey found that most immigrants, even those who have lived in Germany for many years, remain poorly integrated into German society and often face widespread racism and discrimination.[29] This is especially true in what used to be East Germany where much of the anti-immigrant hostility and violence is centered,[30] and while it appears that recent years have seen an overall decrease in German xenophobia, significant pockets of intolerance remain.[31] Research has shown that higher levels of immigration and ethnic diversity generally result in lower levels of trust, altruism, and community cooperation.[32] Interestingly, that holds true not only between different communities but also within the same communities. So it shouldn't be a complete surprise that with the large and rapid inflow of refugees into Germany, these preexisting tensions and rifts became ever more visible and when the story about the widespread assaults on New Year's Eve began to dominate national news, things came to a head. For many Germans, the news provoked a sense of outrage and antiforeigner sentiment, especially since the reporting often emphasized the Arab or North African descent of the perpetrators. The image of white women being victimized by dark-skinned foreign men served to further inflame anti-immigrant sentiment.[33]

Simply put, on my 2016 trip to Berlin, I was witnessing a profound backlash against these recent arrivals. What was ironic is that western European Union nations have often championed progressive policies and attitudes in regard to human rights in general and toward refugee populations specifically. Moreover, since the 1990s, Europe had been moving toward the removal of borders and walls, and this trend was solidified with the Schengen Agreement, which allowed for free travel within internal European borders.[34] Those values and policies were now being challenged by the deep-seated fears, nationalism, and racism bubbling to the surface in the wake of such a rapid and large influx of refugees. Keep in mind that European economies have been largely stagnant in recent years and this economic malaise added to a fairly widespread sense of unease and economic insecurity. Events such as the terrorist attacks in Paris, Nice, and Brussels in late 2015 and early

2016 further aggravated such prejudice and hostility. Germany also experienced several acts of terrorism and a mass shooting during the summer of 2016, and that they were mostly perpetrated by those from immigrant backgrounds or recent refugees further reinforced and strengthened the growing backlash.[35] During this same time period, other European nations, such as Sweden, Greece, France, Belgium, Austria, and Denmark, also saw a dramatic shift in attitudes toward the influx of refugees.[36] In short order, a groundswell of opposition to the refugee crisis emerged that resulted in anti-immigrant legislation in a number of European Union nations, and far-right political parties making significant electoral gains in nations such as Germany, France, the Netherlands, Greece, Hungary, Sweden, Austria, and Slovakia.[37] All too often these attitudinal and legislative changes also resulted in anti-immigrant violence. Germany, for example, has seen a dramatic uptick in attacks on migrants, rioting, and arson attacks on refugee centers, more than two hundred in the first half of 2015 alone.[38] Similar violence has been recorded in Greece, Sweden, and France among a number of other nations, while in the United States anti-immigrant violence targeting Latinos and Muslims has also increased dramatically.[39] Clearly, border issues and large-scale population movements are volatile sources of tension and conflict. Granted, the limited violence and overt hostility that we've seen in recent years against immigrant populations is still a far cry from genocide in which entire groups are targeted for elimination because they are perceived as dangerous outsiders, but it is deeply worrying nonetheless. It is precisely such attitudes and persecution that, given the right circumstances, can morph into something much more widespread, systematic, and deadly. Of this trend, Christian Parenti writes,

> As the planet warms, the political tumors of American authoritarianism, our current repression of immigrants, will metastasize. A similar illness infects Europe, and climate change will intensify even if necessary mitigation finally begins. Already we see the forms that adaptation in the developed world will take. The de facto authoritarian, cryptoracist state hardening, encapsulated by the war on immigrants, will accelerate as climate-change-driven migration becomes an ever more pressing issue.[40]

Clearly, large-scale population dislocation can facilitate the development of intergroup resentment, hostility, and violence. How will such tensions play out in the coming years as the numbers of refugees grow? Will large-scale population displacement foster violent reactions and responses? While we don't know the answer to these questions, we do know that climate change is expected to dramatically accelerate the issue of forced displacement and increase the number of refugee and internally displaced populations.[41] Even as early as 1990, the first report of the Intergovernmental Panel on Climate Change (IPCC) concluded that "the gravest effects of climate change may be those on human migration."[42] In fairness, the dilemma of uprooted populations is nothing new. History is replete with examples of such displacements, which have often involved conflict and climate-related triggers.[43] What is unprecedented, however, is the scale and scope of the expected displacement. While around 25 million people were displaced during the 1990s due to environmental degradation and because of natural disasters,[44] by the year 2050, some projections suggest that as many as 700 million people may be on the move.[45] The International Organization for Migration takes a more conservative approach and places the probable number at around 200 million.[46] Even with this middle-of-the-road projection, we have to understand just how massive a problem that will present and the amount of suffering, dislocation, and trauma those impersonal numbers truly represent. In the modern era, the largest single displacement of people occurred in Europe after World War II and involved anywhere from 40 million to 66 million people.[47] That considerable movement of people, spawned by war, genocide, deportations, forced labor, death, and destruction led to years of subsequent social unrest, tremendous amounts of violence, civil war, and revolution.[48] Yet those numbers will be dwarfed in the coming years by the size of the population dislocation expected to occur as a consequence of climate change and its attendant impacts on human communities.

So how exactly will climate change contribute to migration? The answer to that question is a bit more complicated than one might think. While classic theories of migration tended to focus on simple push-pull models that suggested economic, environmental, and demographic conditions *push* people out of an area, and things such as available land, economic opportunity, and peace *pull* people into an area,[49] adding climate change into the mix forces us to consider other factors. One important variable concerns what is often termed "vulnerability," which simply refers to the adaptive capacity and resources of a given community or region.[50] In other words, population displacement is not solely determined by the type, amount, and severity of disasters and climate-induced changes, but also by individual and communal capability. How well prepared, in other words, are people and groups to mitigate, cope with, adapt to, and recover from climate-related impacts?[51] Such capacity can involve many things, including not only economic resources but human and social capital as well. Furthermore, vulnerability can be considered not only in terms of the absolute amount of such resources, but also on how they are distributed throughout a nation, region, or community. Vulnerability, therefore, can similarly concern access, infrastructure, and the ability of a state to deliver aid, support, and resources to those most in need.[52] With fewer such resources and skills available, communities have a reduced capacity to cope with altered climatic conditions. In a very real sense then, migration can be understood as an adjustment to an altered environment.[53] In practical terms, the issue of vulnerability also means that the poorest nations and the poorest communities will generally be those least able to adapt and cope since they possess the fewest economic, social, and political resources. It is a terrible irony of the global landscape that those least responsible for climate change will be most affected by it. Large segments of the developing world, mired in widespread poverty, largely dependent on agriculture, prone to political corruption and violent conflict, have relatively low levels of resiliency and are the most vulnerable to climate shocks. It is highly likely that portions of their populations will represent the majority of those displaced, given that migration will be one of the relatively few options available when catastrophe hits.

At times, groups and individuals will be forced from their homes and communities by natural disasters such as storms and floods.[54] Tornados, cyclones, hurricanes, and other extreme weather events are expected to increase in frequency and severity and will result in sudden disruptions and relocations. Often referred to as distress or forced migration, such movements tend to be temporary, with those who have left often returning as soon as possible. Alexander Betts refers to such displacement as survival migration.[55] This kind of population movement, because of its temporary nature, also tends to be largely internal rather than international in scope.[56] Even though we tend to focus on international immigration, the vast majority of displaced people usually stay within the borders of their own nation. They may relocate to refugee camps in another part of the country or simply move from increasingly unlivable and impoverished rural communities into cities where more opportunities and resources exist, although all too often they have merely traded rural deprivation for urban poverty.[57]

Other times the causes for population shifts will arrive more gradually in the form of chronic problems such as drought and less reliable monsoon rains. Such incremental changes will over time decrease the agricultural yield of a region and impair the ability of that area to sustain life. Keep in mind that agriculture is the source of survival for about 85 percent of rural people and in some locations, such as Africa, over 80 percent of the population is rural, 90 percent of whom are subsistence-level farmers.[58] Issues such as environmental degradation, increasing levels of soil salinity, and desertification may have comparable effects in terms of progressively forcing those in the impacted areas to leave for better regions. At first, the gradual accumulation of such problems will likely result in increased numbers of migrants who are simply following preexisting seasonal and labor migration patterns.[59] While such movement is usually temporary, over time the seasonal nature of such movement may become more and more permanent for many of those making the journey. Ironically, another potential source of migration may involve communities and populations being forced to move by their own government because of climate mitigation efforts such as expanding forest areas to serve as carbon sinks or damning rivers to create reservoirs to cope with water shortages.[60] For communities in the way of such amelioration projects, displacement may be dictated by state policy according to coldly utilitarian calculations.

In addition to all of these causes noted above, another important source of population displacement will be wars and genocide. Such violent conflict can easily spur an exodus from a region as populations attempt to escape death and destruction through migration.[61] Displacement also often precedes violent outbreaks because of the heightened ethnic and other intergroup tensions that can lead to scapegoating and repression even before the onset of large-scale violence.[62] Research demonstrates that political violence is the most significant cause of large-scale population displacement.[63] Furthermore, most conflict-based forced displacement arises out of only a select few types of collective violence, namely, genocide, politicide, and civil war.[64] Sometimes forced displacement is not merely a by-product of war and genocide, but actually constitutes a central strategy of the violence. In Bosnia, for example, Bosnian Serb paramilitaries and army forces sought to ethnically cleanse territory through torture, sexual assault, and mass murder. Those they didn't lock up in concentration camps or kill outright were terrorized into fleeing into Bosnian Muslim territory.[65] Similarly, in the Sudan, the forcible displacement of members of African tribes from their farms and communities by Janjaweed militia and government forces was a central goal of the genocide, or as the researcher Jérôme Tubiana explicitly asserts, "The massive displacement is not merely a consequence of the attacks, but rather a central war aim of the attackers, who are clearing entire areas of their inhabitants."[66] As we discussed in the previous chapter, such tactics were pursued in order to provide more access to water and grazing lands for the seminomadic Arabic tribespeople at the expense of the more African and agricultural tribes who were being forced out through violence and terror.

The connection between violent conflict and population migration can also operate in reverse order. At times, forced displacement is not an outcome of conflict but is instead one of the root causes. In his extensive analysis of conflict and migration, Myron Weiner examined the relationship between war, political repression, and migration patterns and found that a ripple effect sometimes occurs. Specifically, his analysis revealed that when conflicts erupt that displace populations into neighboring regions and states, such a movement of people often serves to destabilize the entire "neighborhood," to borrow his terminology, and thus sparks more conflict and more refugees. He notes, for example, that the war between India and Pakistan in 1972 was caused by the ten million refugees from East Pakistan that had fled the civil war during the previous year.[67] In a similar vein, the work of Idean Salehyan and Kristian Gleditsch reveals that when refugees fleeing violence dispersed to surrounding countries, they increased the risk of civil war within the host nation. They did so by facilitating the flow of weapons, combatants, and ideologies supporting violence across borders; by changing the ethnic composition of the host nation; and by heightening competition for resources and jobs.[68] Refugees, in other words, can spread the seeds of violent conflict as they travel outward from the source of fighting.

In his assessment of climate change and migration, Rafael Reuveny suggests four primary channels or mechanisms that contribute to the likelihood of violent conflict arising out of migration.[69] The first, he notes, is competition. In some situations, refugees will place an additional load on resources and such increased pressure on the carrying capacity of a region can promote competition for access and distribution of those resources between the newcomers and those native to that area. If vital resources are scarce or in danger of being depleted, such conflict is intensified. Furthermore, the increased burden on local supplies can create pressure to acquire needed supplies from neighboring regions and thus increase the risk of conflict being displaced geographically. Reuveny's second channel is ethnic tension and refers to a situation in which recent arrivals are from different ethnic groups. Such an influx can foster feelings of resentment and fear on the part of native populations because the changing ethnic distribution is perceived as threatening the preexisting balance of power. Distrust is the third pathway and is closely related to the second one. Essentially, increased feelings of suspicion and hostility can develop between the host nation and the country of origin. The fourth channel, according to Reuveny, is what he terms fault lines and denotes conflict and competition arising along socioeconomic fault lines of economic distinctions rather than ethnic ones, such as when recent migrants compete with native workers over jobs. Reuveny's analysis helps us understand that the rapid influx of large numbers of migrants and refugees is often quite difficult for host societies to manage and even well-resourced and politically stable nations can struggle to absorb significant numbers of new arrivals, and the strains of doing so may well heighten intergroup tension. The refugee crisis in Europe and the resulting backlash against recent immigrants certainly underscore that reality.

Immigrant groups can easily become the focal point of hostility, especially if the new arrivals are visibly identifiable as being different or are portrayed as representing some sort of danger. That perceived threat might be as simple as believing that the recent arrivals are placing an undue burden on diminishing resources or as complicated as believing that the new arrivals pose an existential threat because of historic enmity and prejudice. The past and present are full of examples of politicians, social elites, bigots, and demagogues manipulating and exploiting xenophobia, fear, anger, and prejudice. During times of instability and threat, such beliefs gain a new potency. Why do social and political leaders stir up and rely on such virulent sentiments? At times, they have done so because they are prey to the same prejudicial beliefs and values they exploit, while in other instances, such tactics are cynical manipulations intended to gain or retain power by heightening in-group solidarity and mobilizing a population against a manufactured common enemy. These same leaders also often scapegoat groups and hold them responsible for real or imagined ills in an effort to divert attention away from failed policies. In locations where long-standing antipathies and prejudice have wide currency and appeal, such processes of scapegoating are relatively easily accomplished since the process builds on existing stereotypes and beliefs.

In the same way that Hitler and the Nazis scapegoated the Jews for Germany's loss in World War I and for the breakdown of Germany's economy during the Great Depression, so too may leaders in an era of climate change seek to blame vulnerable groups for the problems of that particular place and time. Unfortunately, prejudice and intolerance are not easily eradicated and can be revitalized given the right circumstances and political, religious, and social leaders willing to exploit them. Furthermore, we need to acknowledge that refugee groups are also incredibly vulnerable populations. Legal protections are limited, and often ignored. Because they are stateless, refugees are highly dependent on international organizations and NGOs to protect and provide for them and such aid has often been lacking.[70] Refugee populations also tend to be unorganized and fragmented and such populations are particularly exposed to certain forms of violence such as genocide, which, if you remember, is largely about victimizing defenseless populations.

One particular area where we will see the issue of migration and refugee populations play out in stark relief are at the margins of nations and regions. In a recent interview, author Todd Miller made this point when he stated: "As we enter an era of extreme and potentially catastrophic weather, unprecedented in human history, there will be more militarized and violent international divides that refugees and displaced persons will have to face. These border zones will be the terrain of battles between the rich and the poor."[71] In many ways, the issue of forced displacement is also a question of borders. It is in such regions that the politics of migration are heightened because they are the regions in which migration is often first seen. Refugee camps, for example, are often situated within the frontier zones of nations and this is also where the bulk of new arrivals tend to be concentrated, at least initially.[72] Ironically, refugees are not a product of a failure of the international system of sovereign states, but are instead a consequence of such a system since ideas of territorial integrity are intrinsically built upon conceptions of political boundaries—borders in other words.[73] Borders are also symbolic as well. The U.S.-Mexico border, as we've discussed earlier in this chapter, has become a highly charged and polarizing issue around which questions of immigration, violence, criminality, and identity increasingly revolve. This "crimmigration," as a number of criminologists[74] have termed this process whereby immigration becomes criminalized, is built upon constructed narratives depicting the border as a war zone in which Border Patrol agents combat terrorists, drug mules, and gangs of violent offenders, all intent on infiltrating our nation in order to exploit, victimize, and harm our society. Political and social leaders talk about hardening borders, building walls, and otherwise cutting off the flow of people into the United States. The writer and photographer Krista Schlyer captures this dynamic beautifully when she writes:

> Tragedy has become a defining characteristic of the twenty-first century borderlands: not simply the tragedy of failed governmental policies triggering drug wars, overwhelming human poverty and migration, and cultural and environmental degradation on a massive scale; but also, the great tragedy of misinformation that has drained the public of the will to demand real solutions for the region.[75]

In Europe, we have seen similar sentiments and actions arise in the wake of the refugee crisis of the last few years. Europe has seen a massive investment in reinforcing borders with physical barriers, military and law enforcement patrols, and surveillance technology to such an extent that one journalist refers to it as "the most sustained and extensive border enforcement programme in history."[76] Keep in mind that the militarization of Europe's external borders and the U.S.-Mexico border have not happened in response to a military threat, but are instead entirely aimed at preventing civilian populations from crossing. Ironically, it was in Europe with the founding of the European Union that the notion of borderless states first took hold with free and unrestricted travel between member states, but which are now re-trenching from such a position in order to more effectively prevent refugees from crossing into their territory.

Often defined as frontiers, border areas are usually on the periphery of state power and are often more lawless and less well regulated than regions closer to the core. Because they are physically and culturally more removed from the centers of power and because state control is often weaker in such areas, a frontier mentality can dominate these liminal regions in ways that allow and facilitate violent conflict. In loca-tions where border regions experience a heightened influx of those fleeing conflict or deprivation, it's easy to imagine scenarios in which recent arrivals are subjected to persecution and violence. While distinct and definitive on maps, borders are often much messier on the ground with people, goods, and services flowing back and forth across frontiers and creating areas of interaction and intermingling. Borders, as one scholar noted, "generally bring people together at the same time that they separate them."[77] In fact, the concept of borders is as much about "bordering" as it is about geographic boundaries ("bordering" referring to the transactional processes by which boundaries, identities, and rela-tionships are negotiated, defined, and understood).[78] The geographic confluence of national, religious, and social groups in frontier regions situates states, diverse populations, and variable resources in close prox-imity to each other. All too frequently such a convergence has resulted in friction, hostility, and competing claims for territory, resources, and populations. Although it goes against prevailing wisdom, living in close proximity to other groups does not necessarily foster greater acceptance and harmony, but instead may just as likely result in violent conflict.[79] By definition border zones are home to majority and minority groups and in ordinary periods these populations may coexist peacefully, but during more difficult times, such as during wars, economic downturns, or periods of political instability, any underlying intergroup tension can assume a new and more potent significance. Old prejudices, past con-flicts, and intergroup rivalry can enable various forms of communal violence.

In his book *The Age of Triage*, Richard L. Rubenstein wrote that the modern era is a time of "mass surplus population," which he defined as "a surplus or redundant population . . . that for any reason can find no viable role in the society in which it is domiciled."[80] It is also, he further argued, an age in which whole groups of people are sorted into categories based on their quality or utility. Such thinking is captured by a saying from the genocidal Khmer Rouge regime: "To keep you is no profit, to destroy you is no loss."[81] As societies around the world are confronted with scarce resources and increasing numbers of refugees, is it possible these new arrivals may be defined as surplus? Will they be seen as having no value because they are defined as being different and not useful or necessary? If so, such a scenario paints a troubling picture of potential violence and genocide. We've seen such attitudes in the past. The genocidal impulse has often been fueled by prejudice and xenophobia that portrayed entire groups as being expendable, useless, and a drain on resources. Of this issue, Harald Walzer bleakly predicts:

> All the historical evidence makes it highly probable that "superfluous" people who seem to threaten those already enjoying relative prosperity and security will lose their lives in increasingly large numbers, whether from lack of food and clean water, from frontier wars, or civil wars and interstate conflicts resulting from changed environmental conditions. This is not a normative statement; it simply corresponds to what has been learned from solutions to perceived problems in the twentieth century.[82]

Where we live, geographically speaking, helps shape and influence our individual and communal identities, how we live, and what we believe and value. Such spatiality has meant that border zones are regions that have often been swayed by belief systems and popular movements defining specific population groups as a source of contamination and pollution and deciding to remove the offensive physical presence of the group so defined. This is the essence of genocide: the ostensible purification of a nation through the physical elimination of a communal presence that, because of ethnic, political, religious, or other differences, is considered a threat. Border zones tend to be more demographically diverse, and as we've discussed above, are prone to a great deal of intermingling and interaction. Because of this, they also tend to be seen as particularly dangerous sources of racial, political, ethnic, religious, or other forms of pollution and any examination of past examples of communal violence, such as genocide and pogroms, reveals that they have often been perpetrated in border regions. In his examination of genocide in Eastern Europe—what he refers to as the European Rimlands—the British historian and genocide scholar Mark Levene found a lethal combination of failing empires and attempts to maintain and assert control over territory and populations often resulting in violence and persecution as struggles over national and ethnic identity intensified, multiethnic communities were destabilized, and territory was contested.[83] There's a reason why historian Timothy Snyder coined the term "bloodlands" to refer to this section of Europe where borders abound, multiethnic populations predominate, and the politics of empire, race, and nationalism have resulted in frequent conflict, war, and genocide.[84]

We should also note that borders change over time. Far from being static and unchanging, political borders have often been remarkably fluid or "plastic" to borrow Michel Agier's terminology.[85] As political and territorial fortunes have waxed and waned, so too have boundaries shifted in response. But historic claims to land are not easily or quickly forgotten and as states confront an altered environment with diminishing resources, and as they eye the resources of neighboring countries, the solutions imagined may well be informed by ideas of historic territorial loss. War and genocide may become tools not only to acquire needed resources but also to reclaim lost land. Such territorial losses and perceptions of unjust victimization have often played a powerful role in shaping national identity and, as we've discussed previously, these ideas can be very important in helping facilitate violence, whether genocidal or not. The situation in Ukraine provides an object lesson in such thinking.

Ukraine has long been defined by Russia as being essential to its national security and within its particular sphere of influence. Interestingly enough, the word "Ukraine" actually means borderland and the history of this land bears this out. Located as it is in the borderlands or "bloodlands" of eastern Europe, Ukraine has been conquered and subjugated many times over its long history. More recently, as Russia has sought to reassert itself as a world power, control of Ukraine has been understood as being central to Russia's strategic interests. Such thinking was given new urgency by the Ukrainian government's recent moves to distance itself from Russia and develop stronger ties with the West. We also shouldn't discount the role that access to Ukraine's abundant natural resources has played in creating a situation in which Russia has fomented and supported a separatist movement in Ukraine that has turned portions of this nation into a war zone.[86] In short, the current situation in Ukraine speaks to the ways in which border issues can play out in dangerous and violent ways. Border regions, in short, are intimately linked with climate change, conflict, and genocide on a number of different levels. The physical and social terrain of such regions has made them vulnerable to conditions and ideologies that facilitate violent confrontation and conflict. When we combine the issue of borders and population displacement, one nation stands out in terms of predicted risk, and that is Bangladesh.

Bangladesh is the seventh-most-populated country in the world with more than 150 million people. Located at the confluence of the Ganges, Brahmaputra, and Meghna river systems, almost the entire country is little more than a river delta composed of the sediment from these three rivers and the tributaries that feed them. Because of its geography, this low-lying and densely inhabited nation is particularly vulnerable to the twin issues of flooding and forced migration as a consequence of the former. In fact, Bangladesh sits on the largest floodplain in the world. Most of the nation is only around one to three meters in elevation and only in the north and east does the nation have any hills of significance.[87] Throughout the rest of the country, lowlands interspersed with creeks and rivers are the norm. Because of this reality, flooding is a way of life in Bangladesh, whether through episodic events such as storms or through the normal annual floods resulting from high rainfall amounts during the monsoon season.[88] The Bay of Bengal is a breeding ground for tropical storms that often funnel north to slam into the coast of India and Bangladesh. Sixteen out of the thirty-five worst tropical storms with death tolls over five thousand impacted Bangladesh.[89] The portion of eastern India that abuts Bangladesh has also suffered significantly, although not quite as badly. The history of this region, and of Bangladesh in particular, is full of storms and floods with widespread death and destruction typically resulting from these inundations. Cyclones coming ashore on the coast of Bangladesh caused 11,069 fatalities in 1985, 5,708 in 1988, and 138,000 in 1991.[90] These numbers are staggering when one considers the amount of suffering and trauma they reveal. In addition to the large numbers of people killed, these storms and the attendant flooding also resulted in hundreds of thousands being made homeless. Significant losses of crops also accompanied these catastrophes. One storm in 2016, for example, resulted in over 500,000 coastal Bangladeshis being evacuated, with many of them subsequently being left homeless.[91]

In the coming decades, this kind of situation can be expected to occur more frequently since, as we've seen, one important consequence of climate change will be more frequent and more severe extreme weather events. Over time, such temporary flooding will be accompanied by permanent loss of land as rising sea levels make significant inroads into the low-lying terrain of this delta nation. In a cruel illustration of the ways in which climate change impacts will overlap and compound each other, Bangladesh, when not experiencing floods, will also be dealing with the effects of drought. Projected changes to monsoonal patterns over the entire Indian subcontinent will result in less rain during the monsoon season and less snowfall in the Himalayas, both of which will mean that the many rivers Indian and Bangladeshi farmers rely on will experience substantially decreased flows. These natural causes of drought will be compounded by Indian dams further constraining the flow of water into Bangladesh. The two countries share fifty-four rivers and India has increasingly diverted water from those rivers to meet its own agricultural and living needs.[92] Not surprisingly, water-sharing issues have become a frequent source of tension between India and Bangladesh.

The issue of drought has important consequences for Bangladesh's ability to feed and nourish its population. Less water means less wheat and rice can be grown. This is amplified by the fact that over half of all agriculture in Bangladesh depends on irrigation.[93] With less inhabitable land and a diminished carrying capacity for the remainder, life in Bangladesh will become that much more unstable and insecure. The Bangladeshi economy, based largely on agriculture, will also be hard hit by decreased production resulting in the government having fewer resources just when such capital will be most needed to cope with the urgent demands of its people. Summarizing the risks for Bangladesh, one scholar writes:

> Sea-level rise, severe storms, repeated floods, increased water salinity, and worsening water scarcity will directly affect the availability of food. In a country with a large poor population, the decline in food security is a recipe for political and social instability. The decline in food production will make the country dependent on the international food market, the volatility of which needs no elaboration after

the events of 2008. Combined with fragile political institutions, a contentious political culture, and a continued preponderance of violence, the possibility of radicalization is very likely. Militant organizations will make use of these to their benefit. The huge pool of urban poor, particularly new migrants from rural areas, may serve as a reservoir of a disgruntled army.[94]

Given such a situation, it's quite probable that many Bangladeshis will be forced to relocate because their land, homes, and communities have been flooded, because they have no food or jobs, and because they are hoping to escape the societal disorder, extremism, and violence that will surely crop up in the wake of such widespread devastation and dislocation. Many of these displaced people may stay within Bangladesh in regions further inland and higher in elevation. Others, however, will likely seek refuge internationally. Some Bangladeshis may attempt a long, expensive, and potentially dangerous journey to nations in the developed world, but it is also extremely probable that many will look for shelter and safety in neighboring states, the primary recipient, because of a shared common border, most likely being India.

Already struggling to meet the challenges of climate change within its own borders, how will Indian society react to the influx of Bangladeshi neighbors? The former U.S. director of national intelligence provided a briefing to Congress in 2010 in which he said: "For India, our research indicates the practical effects of climate change will be manageable by New Delhi through 2030. Beyond 2030, India's ability to cope will be reduced by declining agricultural productivity, decreasing water supplies, and increasing pressures from cross-border migration into the country."[95] Will increased levels of hostility develop in response to this influx of destitute refugees, especially given the history of animosity between these groups?[96] In recent years, India has already seen anti-immigrant riots resulting in hundreds of deaths, most of the victims being Indian nationals moving within the country in search of economic opportunity.[97] We also shouldn't forget that politics and nationalism, compounded by religious differences, resulted in a great deal of communal sectarian violence around the founding of Pakistan and East Pakistan with periodic flare-ups continuing to the present day. In 1947, when partition created Pakistan and East Pakistan (now called Bangladesh) along religious lines, ethnoreligious violence killed around one million people and resulted in fifteen million forcibly displaced in what sociologist Abram De Swaan refers to as a megapogrom.[98] More recent-ly, competition in the 1980s over farmland in the Indian states of Assam and Tripura between immigrants from Bangladesh and native Indians resulted in ethnic clashes and violence. In one instance, a riot in 1983 resulted in around 1,700 Bangladeshi immigrants being killed.[99]

Bangladesh is predominantly Muslim, while India is largely Hindu. Ethnic tensions and violence already abound throughout rural India, and in such a climate of fear, anxiety, and social disruption these past and present antagonisms could be refocused and repurposed to deal with the new threat of Bangladeshi refugees.[100] Such a potential reac-tion is not that far fetched when we consider that India has already begun construction of a 3,200 kilometer fence on the Bangladeshi bor-der that is expected to be finished sometime in 2017.[101] In the words of one journalist, "as climate change accelerates, the fence will only in-crease cross-border tension. . . . If the Asian monsoon becomes harsher and sea levels continue to rise, the fence in its current form won't be enough to keep Bangladeshis from fleeing to India. But the anti-immi-grant mindset behind the fence might make matters worse."[102] In India, members of the Hindu far right have also begun advocating for mass deportations of Bangladeshis presently living in India.[103] If preexisting patterns of migration from Bangladesh into India have already exacer-bated religious and political nationalism within both communities, what will we see when flooding and drought dramatically increase the pres-sures for Bangladeshis to flee into neighboring India? If previous histo-ry between Indians and Bangladeshis is any indication, this is a situation that is fraught with tremendous risk for persecution, violence, and gen-ocide.

In summary, all the indications strongly suggest that for the various reasons discussed above, climate change will displace many millions of people around the world. As entire populations migrate in search of stability, safety, and opportunity, they will face many dangers, not only in terms of the journeys they undertake, but also in terms of the recep-tion many may find in the host countries where they seek refuge and shelter. Will nations welcome and work to incorporate these newcom-ers into the social, political, and economic fabric of their societies, or will they instead greet the new arrivals with hostility, resentment, and anger?

6

PREVENTING CONFLICT AND BUILDING RESILIENCE

Our goal cannot be to hold climate static. We must understand its menacing and manic moods—and adapt as nimbly as we can to changes in whatever directions and at whatever rates they arrive.

—E. Kirsten Peters[1]

Because the truth is that humans are marvelously resilient, capable of adapting to all manner of setbacks. We are built to survive, gifted with adrenaline and embedded with multiple biological redundancies that allow us the luxury of second, third, and fourth chances. So are our oceans. So is the atmosphere. But surviving is not the same as thriving, not the same as living well. . . . Just because biology is full of generosity does not mean its forgiveness is limitless. With proper care, we stretch and bend amazingly well. But we break too—our individual bodies, as well as the communities and ecosystems that support us.

—Naomi Klein[2]

The evidence is overwhelming and clear; climate change is here. No matter what we do at this point, a warmer world is inevitable. Even if the world were to act immediately and stringently to curtail greenhouse gas emissions, climate change would continue for the foreseeable future, operating as it does on a time delay. The changes we are experiencing in the present are actually the result of greenhouse gases pumped into the atmosphere years ago, while the consequences of today's output won't be felt for years or even decades. The processes of change we have set in motion cannot simply be stopped and reversed. Not only is climate change unavoidable, but it's already making its presence felt. Global levels of carbon dioxide have passed 400 parts per million, a highly worrying and symbolic threshold,[3] while the years 2011 to 2016 have been the warmest on record with each successive year hotter than the last. In 2016 the temperature was .2 degrees Celsius warmer than 2015 and about 1.3 degrees Celsius warmer than preindustrial times.[4] Our planet is already warmer than it has been for over a hundred thousand years. Such high temperatures were last seen during the Eemian interglacial era, which ended about 115,000 years ago and which was notable for much less ice and sea levels about twenty to thirty feet higher than they are today.[5] In recent years, once-rare weather events have become ever more common as drought, flooding, desertification, heat waves, more extreme storms, and many other climate-related impacts increasingly affect communities around the world. Slowly, ever so slowly, humanity is collectively waking up to the reality of these changes and the forced recognition that we must start taking this issue seriously.

This chapter is not about reviewing specific climate change mitigation efforts. Finding ways to reduce greenhouse gas emissions, develop more carbon sinks, reduce dependence on fossil fuels, and geoengineering solutions to the problems of climate change are all strategies that have been proposed to slow or even reverse the rate of climate change.[6] Technology got us into this situation and many believe that technology will get us out of it. We often think that every problem has a high-tech solution,[7] yet expecting and hoping for a technological or scientific fix without also addressing the underlying attitudes and values that created the situation in the first place seem doomed to failure in the long run. Humanity not only has to wean itself from fossil fuel dependency and highly extractive economic systems, but must also develop more sustainable lifestyles, mind-sets, and cultures. As important as these issues are, this chapter is not focused on such technological strategies and solutions. Books have been written on such issues by those far more versed and qualified in such matters and no single chapter could possibly do justice to these specific measures. It is also not my goal to attempt to comprehensively address all of the potential solutions for reducing collective forms of violence, especially genocide. Scholars and activists have long struggled with identifying and developing effective strategies to prevent or intervene in genocide.[8] While many of these proposals have merit and potential, to attempt an inclusive review of these approaches is simply beyond the scope of this single chapter. Instead, my goals are much more basic. In this concluding chapter, I merely want to offer the reader some thoughts and considerations to help make sense of the risks posed by the coming years and the possible strategies that may be designed and implemented to reduce or prevent such risks from manifesting in collective violence. It is by no means intended to be definitive or comprehensive, but rather a way to highlight some key ideas, concepts, and limitations.

Thus far in this book, I have sought to illustrate the various ways in which violent conflict can emerge both directly and indirectly out of the projected impacts of climate change. Specifically, I have suggested that climate change will facilitate the development of various structural, ideological, and psychological conditions that escalate the risk of large-scale organized violence as both an expressive reaction and an instrumental solution to the challenges created by a changing environment. In various shapes and permutations, these will include riots, pogroms, civil and international wars, genocide, and related forms of mass human rights violations. While many of these violent conflicts may come to involve ethnic, national, religious, or political dimensions, underneath such overt qualities, however, they will have developed, at least in part, because of climate-change-induced crises and stressors. In such highly charged situations, preexisting conflicts may take on an entirely new significance as the stakes increase, while in other cases, preemptive wars may be triggered as governments anticipate an attack on them for territory or resources and choose to respond proactively. Another potential pathway may involve states weakening and failing, allowing for the rise of widespread criminal violence, warlords, organized crime, revolutions, civil war, and genocide. State failure may also be an outcome of resource scarcity with violent solutions employed as a means of propping up failing regimes, as a way of protecting scarce and diminishing resources, or as a tool to acquire new ones. As we've discussed in previous chapters, such examples of state failure and the ensuing violence will not be limited to the nation undergoing such a breakdown; they can easily spread throughout an entire region as refugees flee in search of safety and opportunity and, in the process, foster societal instability as their arrival strains the resources of neighboring states and increases intergroup tension and hostility. Importantly, many of these potential outcomes will operate in a feedback loop with each reinforcing and amplifying the others. States, for instance, may be weakened by diminishing natural resources that reduce revenue streams and hinder the ability of a state to confront climate-induced situations. This inability to meet the needs of a citizenry could then heighten levels of social disorder and violence, which, in turn, may serve to further weaken the state and its ability to control its territory and protect its people.

I've further argued that in the stressful and challenging conditions arising from an altered environment, populations will become much more receptive to punitive attitudes and policies that increase the risk of repression and various forms of collective violence against groups perceived to constitute a threat or as a drain on valuable resources. During difficult times, states and political leaders tend to become more repressive and authoritarian in orientation. Old prejudices and xenophobic beliefs will gain a new resonance in settings where a history of previous intergroup hostility and conflict provide a convenient focus for fear, anger, and frustration. Unscrupulous political, social, and religious leaders can tap into such reservoirs of intolerance and stoke them into white hot flames that can then be used to pursue political, social, and religious agendas through scapegoating vulnerable populations and implementing persecutory policies. For all these reasons and more, the coming years and decades will likely see a significant increase in the risk of various forms of collective violence as the changes and stressors mount in communities and nations around the world and as traditional responses are overwhelmed or fail to meet the new challenges as resources draw down. Such violent outcomes will come at a tremendous cost in individual and collective human suffering that should not be forgotten, ignored, or minimized. When looking at the potential consequences of climate change conflict, it is important to remember that the total possible numbers of dead, injured, or displaced represent individual human lives brutally and violently cut short, or forcibly removed from everything and everyone they've known. It's too easy to forget this, however, and so it might be worth reviewing the human costs in more detail.

Group forms of violence have always exacted a staggering toll of death and destruction. If we examine mortality statistics from the twentieth century, for example, we find that according to one estimate, the wars and conflicts from that era alone resulted in around 231 million people killed, a figure that includes deaths as a result of fighting and acts of violence, as well as those indirectly brought about by common by-products of large-scale conflict such as forced removal, famine, disease, malnutrition, mass rape, imprisonment, and similar kinds of consequences.[9] Arriving at a slightly smaller total, political scientist Rudy Rummel, who pioneered the concept of democide, or governmental mass murder, suggested:

> In total, during the first eighty-eight years of this century, almost 170 million men, women, and children have been shot, beaten, tortured, knifed, burned, starved, frozen, crushed, or worked to death; buried alive, drowned, hung, bombed, or killed in any other of the myriad ways governments have inflicted death on unarmed, helpless citizens and foreigners.[10]

World War II alone, for example, resulted in an estimated 65 to 75 million fatalities,[11] while the Soviet Union killed almost 62 million people during its years in power through war, genocide, forced dislocation, planned famines, and the concentration camps of the Gulag.[12] The list of twentieth-century death and destruction could go on and on. Unfortunately, it looks as if the twenty-first century has begun very similarly. Eric Reeves, the foremost scholar of the genocide in Sudan, suggests that around 600,000 have been killed and 2.66 million displaced since the violence began in 2003.[13] We can also examine the consequences of the American invasion of Afghanistan in October 2001 and the subsequent assault on Iraq in March 2003. In both locations, while coalition forces were largely successful in achieving their military goals of removing from power the Taliban government in Afghanistan and Saddam Hussein in Iraq, these respective wars also destabilized those nations, which, as a consequence, descended into religious, ethnic, and political strife. In Iraq, for example, it is estimated that while coalition forces suffered 4,804 military fatalities between 2003 and 2012,[14] there were approximately 119,973 to 132,793 civilian fatalities between 2003 and 2013.[15] Once unleashed, collective violence doesn't always die away easily or quickly and these costs are usually borne by civilian populations. This is especially true when we consider genocide.

In genocide, civilians are the primary targets, rather than unintended casualties or, to borrow a military euphemism, collateral damage. Genocide is a war of extermination against defenseless populations. To get a better sense of how populations can be victimized, scholar Hugo Slim suggests seven spheres of civilian suffering that result from mass crimes and violent conflict:

1. Direct personal violence of killing, wounding, and torturing.
2. Atrocities of rape, sexual violence, and sexual exploitation.
3. Geographic or spatial suffering from forced and restricted movement such as flight, displacement, deportation, destitution, dispersal, resettlement, forced labor, confinement, and detention.
4. Poverty resulting from the first three spheres of suffering with attendant loss of assets and livelihoods.
5. Famine and disease resulting from impoverishment and destitution.
6. Emotional suffering caused by the loss, fear, pain, indignity, and separation from loved ones involved in previous five spheres.
7. Postwar suffering after war is over as a result of landmines, bereavement, missing family members, the challenge of adapting to new circumstances, and losses resulting from missed opportunities such as education and medical care.[16]

These seven areas of civilian victimization reveal that victimization can occur directly from violence and indirectly via a number of common byproducts of violent conflict. His analysis also reveals that victimization can extend for long periods of time, even after the conflict has officially stopped, because things such as poverty, disease, and famine can persist for extended periods of time. It's not easy to eradicate these sorts of problems when the social and physical infrastructures of a society have been damaged or destroyed. How do you eliminate disease when sanitation and medical facilities were devastated in the conflict? How can famine be curtailed when the roads are gone, the fields lie fallow, and the farmers have all been killed or have fled? The consequences of collective violence, in other words, can endure and persist among civilian populations for generations. Significantly, these encompass not only physical destruction but emotional and psychological damage as well.

In many ways, riots, pogroms, wars, and genocide can be understood as a type of human-made disaster because they are so catastrophic and destructive and can affect entire populations and communities. Furthermore, the aftermath of both natural and human-made disasters can linger for extended periods of time as populations struggle to rebuild their communities and overcome not only the physical destruction but the emotional and psychological trauma of the catastrophe as well. Research has shown, for example, that individuals who live through disasters often develop a host of maladaptive behavior patterns that can include acute stress reaction, posttraumatic stress disorder, depression, anxiety, and drug and alcohol abuse.[17] This has been found to be true regardless of whether the disaster is a natural event or a human-caused catastrophe such as war, terrorism, or genocide. In fact, some research indicates that human-caused disasters often result in higher amounts of long-term stress and trauma than natural catastrophic events.[18] Furthermore, while trauma was once perceived solely in individual terms, we now know that trauma can also be passed along at a community level.[19] In genocide, for example, victims are chosen simply because they belong to a group that has been targeted for elimination. When a person is victimized by this crime it has nothing to do with their behavior or beliefs and everything to do with their ascribed identity. Writing about the Holocaust, for example, one historian noted that "the Jews were killed not because they had done anything but merely because they existed—because they had been born."[20] This is the essence of genocide: the destruction of social categories, not of individuals. Entire communities are collectively targeted and collectively victimized. This means that even if specific individuals survive and escape direct victimization, they nonetheless belong to a category of people targeted for destruction and thus often experience a kind of indirect or secondary victimization through that membership.

In many ways, genocide creates a sense of collective victimization among groups that are the focus of genocidal violence. The damage and distress of traumatic events can be transmitted to subsequent generations either *intergenerationally* (within a family unit) or *transgenerationally* (within a population group).[21] The first refers to the ways in which individual families can transmit the experience of victimization through the stories and experiences of their members, while the second refers to the larger community, some of whose members and families may have been relatively untouched by the traumatic event but who nonetheless experience a vicarious victimization due to their membership in the victimized group. Sometimes called historical trauma, this concept can be defined as the "cumulative emotional and psychological wounding across generations, including the lifespan, which emanates from massive group trauma."[22] Even though individual identity tends to be emphasized in Western societies such as the United States, we cannot ignore the reality that collective identity is a powerful part of how people classify and define themselves. Because humans are communal in orientation, our group identities—whether based on nationality, religion, ethnicity, political affiliation, or any other social category—are a central part of our self-identity. James Waller makes this point when he writes, "Group-based identity—whether centered on race, ethnicity, tribe, kin, religion, or nationality—is a central and defining characteristic of one's personal identity and overshadows the self."[23] In many ways, this issue of multigenerational trauma concerns posttraumatic stress disorder (PTSD), which encompasses a range of dysfunctions, both physical or psychological, that can develop after a person experiences something very disturbing and/or harmful.[24] For some severely traumatized individuals, these issues can last for many years and have a strong and ongoing impact on that person's quality of life. In recent years, scholars have increasingly examined the cumulative impact of PTSD over multiple generations. This phenomenon was first recognized and examined in regard to the children of Holocaust survivors, who sometimes showed symptoms of being indirect or secondary victims of the Holocaust.[25] Research suggests that Holocaust survivors tend to suffer from PTSD, and the children of these survivors also often exhibit PTSD, albeit at lower levels.[26] While some survivors were able to move past their experiences and create healthy and productive professional and personal lives,[27] others were not so fortunate or resilient. Some survivors, in fact, exhibit lifelong problems with issues of depression, anxiety, and difficult family functioning.[28] Even those who on the surface appeared to be doing well tended to exhibit more problems and disorders when placed under stress.[29]

In short, the issue of collective violence arising from the direct and indirect effects of climate change is a profoundly important one. While it's relatively easy to calculate the economic, environmental, and political costs of violent conflict, it's much harder to quantify human suffering and trauma. Yet we cannot assess the consequences of climate change without also including the human dimension. War and genocide have always come at a significant cost to those nations, communities, and people unfortunate enough to experience such collective violence. In an era of climate change, such outbreaks of violent conflict arising out of a failure to adapt and cope will undoubtedly result in widespread and long-lasting consequences. Death, suffering, trauma, destruction, and displacement will be the penalties paid by potentially millions of our fellow human beings. Given such profound concerns, what, if anything, can be done?

The first thing to understand is that there are no simple solutions or easy answers. Human behavior, especially when it comes to violence, is simply too complex and contingent. This is why the public intellectual Neil deGrasse Tyson recently tweeted, "In science, when human behavior enters the equation, things go nonlinear. That's why Physics is easy and Sociology is hard."[30] Predicting human behavior is problematic, to say the least. Yet we often ignore this reality and suggest simplistic solutions for highly complex situations. The issue of violence prevention is no exception to this general rule. We often assume, for example, that we can simply export democracy as an antidote to warfare and genocide. In the modern era, especially after the end of the Cold War, such thinking gained a fair amount of currency that was reinforced by a dramatic surge in nations transitioning to more democratic systems of governance.[31] After the fall of the Berlin Wall, the influential American scholar Francis Fukuyama wrote *The End of History and the Last Man* in which he theorized that humanity had reached the end of its ideological evolution and that liberal democracy represented the apotheosis of forms of government.[32] In the wake of the terror attacks of September 11, 2001, the notion that democracy should be exported through regime change and military power gained a renewed prominence in American foreign policy.[33] Because democracy tends to be equated with individual and collective freedom, it's often believed to present a solution to many of the world's problems, especially violent conflict. Democracy, it is felt, creates more law-abiding and peaceful nations because individuals and groups enjoy personal freedom and feel empowered, and because it provides nonviolent processes and procedures for resolving disputes and settling differences. Since democracies emphasize the rule of law and participatory governance, they also often enjoy a relatively high sense of legitimacy in the eyes of their citizenry. In short, according to political scientist and theorist John Keane, violence represents the antithesis of democracy's "spirit and substance."[34] But the assumption that democratic states are inherently less violent is problematic for many reasons.

Democracies often have more criminal, communal, and ethnic violence, and fight more civil wars than authoritarian states generally do. Some scholars have even found that elections, so central to democratic processes, often lead to violence.[35] Groups that are empowered through elections, for instance, sometimes face retaliatory violence from groups seeking to reassert control and power lost in the vote. In some cases, elections can even lead to ethnic cleansing.[36] Moreover, while some have argued that genocides can only be perpetrated by authoritarian regimes and not by democracies because of systems of checks and balances on the use of power,[37] the reality is that nations transitioning to democratic rule or young democratic nations are often very fragile and violent places. One analysis found that the odds of state failure for young or transitional democracies are seven times as high as they are for full democracies and authoritarian regimes.[38] Criminologists Gary La-Free and Andromachi Tseloni have conducted research showing that transitional democracies tend to have a great deal of violent crime,[39] while sociologist Michael Mann points out that the phenomenon of ethnic cleansing, closely related in many important ways to genocide, is a largely modern practice often perpetrated by young or transitional democratic states.[40] His argument is that conceptions of national identity, so important to modern democratic forms of governance, are often defined according to ethnicity. This kind of situation can lead to violent conflict when different ethnic groups inhabit the same national territory yet one lays claim to an exclusionary national identity. In short, exporting democracy is not necessarily the quick fix that it is often envisioned to be. While democratic systems of government have many benefits and advantages for its citizens, we cannot make the mistake of thinking that its rewards can be simply, quickly, or easily exported. Certainly, the examples of Afghanistan, Iraq, Libya, Egypt, and Ukraine in the wake of their recent transitions out of authoritarianism speak loudly and clearly to the difficulties in exporting and maintaining democracy.

Much of the current work on preventing violent conflict, especially in terms of war and genocide, has emphasized international legal protections and mechanisms to intervene in situations where violent conflict has broken out and to punish those responsible for perpetrating such actions. This is especially true where genocide is concerned. Recent years, for instance, have seen the creation of a number of initiatives such as the Report of the International Commission on Intervention and State Sovereignty, popularly referred to as the Responsibility to Protect or R2P,[41] the Albright/Cohen Report on Preventing Genocide,[42] and the Mass Atrocity Response Operations (MARO) project handbook designed to provide practical operational guidelines for military intervention when conflicts with large-scale mass atrocity break out.[43] Each of these has as a goal the prevention and punishment of genocide and similar gross human rights violations. The R2P, for example, is concerned with reinforcing the principle that all states have a responsibility to protect not only their own populations, but those of other nations as well, and to intervene when human rights abuses are being perpetrated.[44] Essentially, it is about creating new international norms establishing the primacy of people over sovereignty and persuading states to honor their legal obligations under international law. When nations fail to live up to international norms on human rights and engage in repression, persecution, and/or genocide, then the principle of nonintervention has to be discarded, according to R2P, out of this responsibility to protect.

In addition to these initiatives, we have also seen a proliferation of international criminal justice mechanisms instituted to punish offenders and deter future outbreaks.[45] This is exemplified by the establishment of the International Criminal Court (ICC), the International Criminal Tribunal for the Former Yugoslavia, the International Criminal Tribunal for Rwanda, the Iraqi Special Tribunal, and the Special Court for Sierra Leone among others. These notable examples of international criminal justice have been mirrored at the national level by a great many domestic courts and prosecutions that have increased dramatically since the 1990s.[46] Nations all around the world, in other words, are increasingly prosecuting and punishing various kinds of human rights offenders. The political scientist Kathryn Sikkink refers to this trend as a "justice cascade," and argues that it has fundamentally altered the international political and legal landscape by creating new international norms and expectations around genocide and similar large-scale human rights violations.[47] To these can be added a great many alternative justice processes that are sometimes grouped together under the rubric of transitional justice and that have taken place in many countries trying to transition from conflict and repression into more stability and peace.[48] These include truth and reconciliation commissions, lustration, amnesty, reparations, or some combination of these. Support for such measures is typically based on the belief that they "promote a host of seeming universal goods, including the rule of law, justice, and peace and reconciliation."[49]

All of these punitive and restorative justice processes have been matched by high expectations for what they can accomplish, namely, that they will ensure peace, build the rule of law, and deter future potential offenders.[50] Hans Köchler, for example, summarizes such common sentiments when he writes that "only if the perpetrators of serious war crimes, crimes against humanity, acts of genocide, crimes of aggression, and crimes of international terrorism are prosecuted according to generally agreed upon norms, will global peace and security prevail."[51] Yet for all of the noble intentions and idealism such comments and perceptions reflect, the evidence to date is decidedly less than impressive. While these international legal measures have constituted necessary and important steps in theorizing and institutionalizing strategies and processes designed to prevent genocide and related forms of atrocity crimes, we have to acknowledge that, as important and meaningful as they are, they have not necessarily been very effective to date in terms of deterring potential outbreaks of such collective violence. After assessing the capacity of international and national trials to deter genocide, Maureen Hiebert, political scientist and genocide scholar, concluded, "Empirical evidence at both the international and domestic levels seriously undermines the rational deterrence thesis that trials lead to genocide prevention through general deterrence."[52] A good example of such a failure concerns the massacres perpetrated in July 1995 after the fall of Srebrenica during the genocidal ethnic cleansing campaign in Bosnia.[53] More than eight thousand Bosnian Muslim men and boys were murdered after being turned over to Bosnian Serb forces by United Nations peacekeepers. This crime was perpetrated more than two years after the founding of the International Criminal Tribunal for the Former Yugoslavia. Clearly, the Bosnian Serb army and militias were not deterred from perpetrating the largest mass killing operation since the end of World War II by the establishment of this court. More recent events in Syria, for instance, or South Sudan and Myanmar bear mute testimony to the continued inability of international courts to deter. In some cases, these courts have actually facilitated the continuation of ethnic tension and conflict by fostering a sense of injustice and persecution among those groups in the crosshairs of attention and prosecution.[54] Furthermore, the long-term viability of such mechanisms is also unclear.

In 2016, Gambia, Burundi, and South Africa announced their intention to withdraw from the ICC because of a perceived bias against African nations.[55] Russia, although never having ratified the ICC, subsequently announced that it was withdrawing its signature from the Rome statute that founded the ICC.[56] Such developments call into question the long-term ability of international criminal justice to survive or to function with the support and cooperation of member states. All of this is not to suggest that international criminal justice does not have a role to play in helping keep the world safer in the coming years. Law is a powerful tool that has many purposes and these courts serve important ends by punishing offenders, strengthening the rule of law, ending cultures of impunity, and recognizing and giving voice to the suffering of victim groups. These are real accomplishments and shouldn't be dismissed, but we also need to acknowledge that such mechanisms are not sufficient by themselves to prevent or even lessen the risks of climate-change-induced violent conflict. International criminal justice must be supported and buttressed in ways that serve to reduce the risk of violent conflict from breaking out in the first place. In this regard, one important area concerns the role of poverty and inequality in fostering conditions that heighten the danger of collective violence.

We know that the poorer nations of the world are expected to experience the most unfavorable impacts because of "geographic exposure, reliance on climate-sensitive sectors, low incomes, and weak adaptive capacity."[57] Simply put, some nations are much better equipped to cope and adapt to the challenges of a changing environment because of the resources they have at their disposal. Throughout this book, a common theme has been that the risks of violent conflict developing out of changed environmental circumstances will not be evenly distributed around the world. While no region of the globe is immune to climate change, some areas will be much harder hit than others due in part to the nature and severity of the specific environmental changes, as well as the ability of communities and nations to absorb and confront such shocks, or some combination of these two factors. The second part of this equation—the capability to deal with climate change impacts—will often be a function of resources and wealth, at least partially. The more resources and wealth there are, the more nonviolent options and solutions there will be to ameliorate the consequence of a changing environment. We shouldn't be surprised by this connection. A great deal of criminological research tells us that violence is disproportionately concentrated among the ranks of the poor.[58] If we examine violence in the United States, for example, we find that while the middle and upper classes can and do become offenders and victims, their perpetration and victimization rates tend to be much lower than they are for those in the lower economic ranks of society. Most violence, in other words, is perpetrated by and against the poor. The criminologist Elijah Anderson captures the dynamics of this relationship when he points out:

> The inclination to violence springs from the circumstances of life among the ghetto poor—the lack of jobs that pay a living wage, limited basic public services (police response in emergencies, building maintenance, trash pickup, lighting, and other services that middle-class neighborhoods take for granted), the stigma of race, the fallout from rampant drug use and drug trafficking, and the resulting alienation and absence of hope for the future. Simply living in such an environment places young people at special risk of falling victim to aggressive behavior.[59]

Cross-national research tells us that the same patterns emerge in other nations as well; the poor are overrepresented among the ranks of those who perpetrate and who are victimized by various violent crimes. But violence is not just a function of absolute poverty; it is also a consequence of inequality.[60] The greater the disparity in a society between the haves and have-nots, the more violence you tend to get. Keep in mind, however, that inequality is not simply a measure of income but encompasses a broader range of resources that include access to health care, education, economic opportunity, the essentials of life, a voice in governance, and so forth. Inequality, as criminologist Elliott Currie suggests, is a form of social exclusion or disadvantage.[61] In short, poverty and inequality are key issues that strongly influence the development of aggressive and violent interpersonal behavior.

   This same dynamic holds true for collective forms of violence as well. When we examine wars and genocide, we find that the risks are highest for the poorest nations of the world where civil wars, revolutions, and ethnic violence are often widespread and endemic. Furthermore, when inequality plays out along racial or ethnic lines, the risks for violent conflict increase dramatically.[62] When the haves belong to one racial or ethnic group and the have-nots to another, that is when we are most likely to see various forms of communal violence. Given the relationship between poverty, inequality, and violent conflict, any approach to reducing the risks of conflict, in addition to criminal-justice-based approaches, must also incorporate strategies addressing this specific connection. In other words, in order to limit the likelihood of violent conflict developing out of climate-change-induced crisis, rates of poverty and inequality must be decreased. This is precisely the approach advocated by the psychologist James Gilligan who argues:

> It has been shown throughout the world, both internationally and intranationally, that reducing economic inequities not only improves physical health and reduces the rate of death from natural causes far more effectively than doctors, medicines and hospitals; it also decreases the rate of death from both criminal and political violence far more effectively than any system of police forces, prisons, or military interventions ever invented.[63]

In other words, if we are truly going to reduce the risk of violent conflict around the world, at some point we must address the inequality that lies at the root of so many forms of violence. This isn't just "pie in the sky" theorizing. The world has already seen a great deal of progress on this front. From 1990 to 2011, the proportion of the world's population living on less than $1.25 a day dropped by more than 60 percent.[64] Furthermore, infants and children in the developing world are now almost 50 percent more likely to survive into adulthood and almost half are less likely to be undernourished.[65] Financially, the costs to end extreme poverty are actually much less than you might believe. In his book *The End of Poverty*, economist Jeffrey Sachs crunched the numbers and found that it would take about $175 billion a year for around twenty years to end extreme poverty and while this might seem like an astronomical sum, consider that it represents less than 1 percent of the total income of the richest nations of the world.[66] While such an investment might seem unrealistic, especially given the economic realities and entrenched interests of capitalism and globalization, it is no less true that measures intended to remedy the disparities in vulnerability will be absolutely essential to confronting a rapidly warming world and all the attendant consequences arising out of it. When we also factor in the costs of conflict and war, the investment in ending poverty seems much more reasonable. Aside from the costs in terms of human suffering, estimates suggest that the global costs of civil wars alone total more than $100 billion a year.[67]

In other words, the developed world must invest in creating the economic, political, and social infrastructures needed to deal with the consequences of climate change. The international community cannot simply stand by and watch the vulnerable nations of the world fail and descend into violent conflict. We already see a tendency among some of the developed nations of the world to do precisely that. This inclination is certainly evident with the recent backlash against immigrants in many nations and growing support for nationalist movements and hard-line political groups. The fear is that the developed world will adopt a fortress mentality and attempt to wall off the rest of the world by hardening borders and cutting off the flow of migrants and refugees. Yet is such a reality even possible in a world with millions upon millions of people forcibly displaced because of environmental degradation, lack of economic possibilities, and violent conflict? Can any political border truly withstand the pressure of so many desperate and needy people seeking safety and refuge? Remember that the consequences of wars and genocide are not limited to the nation in which such violence develops, but invariably spread into the surrounding regions. Furthermore, given that modern technological lifestyles and living standards are highly dependent upon resources not always found in the developed nations of the world, such a retreat is simply not possible. To survive and acquire vital resources, these fortress nations would either need to negotiate and trade with nations in the developing world or send in the military and take such necessities by force; either option requires engagement in the wider world. In many ways, however, this is as much a moral issue as it is a national security one.

There is no getting around the fact that the developed world is largely responsible for anthropogenic climate change because of industrialization, colonialization, and consumerism. The high living standards enjoyed by a relatively few nations have been made possible by the economic exploitation of much of the rest of the world. It has also fostered a mentality of unlimited consumption that is based on largely unsustainable practices. The great irony is that those nations most responsible for climate change are also those that have the wealth, infrastructure, social capital, and political stability to largely insulate or cope with, at least initially, the worst effects of climate change. Yet these few privileged nations enjoy that position precisely because of the practices that created the situation in the first place. These facts suggest that, for moral and practical reasons, those countries with the capability and assets must work to build resilience among those nations and communities considered to be the most vulnerable. Essentially, investing in the most exposed nations is all about building resilience. Remember, if you will, that resilience refers to the ability of individuals and communities to absorb and recover from obstacles, setbacks, and other challenges. The less resilient a community, the more difficulty it has in confronting and coping with difficulties. Highly resilient states will prove to be more resistant to increased risks or as one U.S. Foreign Service officer recently wrote: "States that have legitimacy across population groups, strong courts, developed institutional planning mechanisms, and strong economies are very likely to be able to manage the effects of climate change before they take on a dynamic with a potential for violence."[68] Resiliency is often, at least partially, a question of resources, but keep in mind that resilience is more than just about material wealth and resources, although that's certainly a big part of it, but also includes human capital. Resiliency is not just about having money and supplies, but is also about nurturing a population that has the education, knowledge, training, ingenuity, and imagination to develop and implement solutions. But how do we build resiliency? What specific strategies and tactics will be most effective? How do we invest in vulnerable nations and communities without replicating paternalistic and exploitative mind-sets that were so critical to enabling past colonial practices and injustice? How do we overcome political and social resistance in order to implement such steps? Will we rise to the challenges or will we descend into a Hobbesian nightmare of everyone for themselves and never-ending resource wars? These are incredibly difficult questions to answer. But try we must and we must do so swiftly. Time is increasingly in short supply. This doesn't mean, however, that we should give up hope. Far from it.

In many ways, this book has been an exercise in examining potential outcomes and worst-case scenarios, and they are frightening indeed. The possibilities of state failure, resource scarcity, food and water shortages, and mass population displacement leading to xenophobic hostility and persecution, riots, pogroms, civil war, and genocide are very real potential consequences. But we need to remember that we live in a world of probabilities, not certainties. Such outcomes are not preordained. Violence, both individual and collective, is highly contingent; it is never inevitable. The specific ways in which climate change plays out in the coming years will depend a great deal on the precise impacts in different locations and the localized, regional, national, and international reactions, choices, and calculations made by individuals, communities, nations, and their leaders. Violence, we must remind ourselves, is not the only response possible during difficult times. Communities can come together to support, assist, and collaborate in seeking solutions to problems. In his recent book *Tribe*, writer Sebastian Junger points out that research has shown that sometimes during times of crisis or disaster—such as the London Blitz or in the immediate aftermath of earthquakes—communities have sometimes rallied together to help and support each other despite previous differences.[69] Referring to the work of researcher Charles Fritz, who studied such events, Junger writes:

> Disasters . . . create a "community of sufferers" that allows individuals to experience an immensely reassuring connection to others. As people come together to face an existential threat, Fritz found, class differences are temporarily erased, income disparities become irrelevant, race is overlooked, and individuals are assessed simply by what they are willing to do for a group.[70]

In other words, rioting, war, and genocide are not always the preferred methods of response or the intolerant reactions of communities in crisis. Crisis can bring out the best in people. Furthermore, we also shouldn't forget that for all of its violent conflicts and problems, the world today is a safer and less violent place than it has ever been before. One scholar recently noted this when he pointed out that

> from a historical perspective, it seems evident that humankind is actually enjoying the most peaceful and prosperous era ever. In the early twenty-first century, for the first time in history, more people

die from eating too much than from eating too little; more people die from old age than from epidemics; and more people commit suicide than are killed by war, crime, and terrorism put together.[71]

This assertion is mirrored in the work of the well-known psychologist Steven Pinker, who, in his book *The Better Angels of Our Nature*, argues that the current era has seen a dramatic decrease in war, genocide, slavery, rape, hate crimes, riots, and other forms of violence, both individual and collective.[72] Similarly, historian Ian Morris argues that in the long run, war, as destructive and terrible as it is, has made the world a safer place.[73] It has done so by strengthening states and state institutions that have become more capable at reducing internal amounts of criminal and other violence. The rewards and costs of warfare have also changed so that international wars are now a relatively rare event. The National Intelligence Council made this point when they wrote, "Historical trends during the past two decades show fewer major armed conflicts and, where conflicts remain, fewer civilian and military casualties than in previous decades."[74] The purpose of pointing out such trends is not intended to diminish or minimize the suffering of those who have experienced the tragedy and horror of war or the indignity of poverty and victimization firsthand, but rather to highlight the fact that the world today, in general, is a much safer and more prosperous place today than it was in the past. This is encouraging, to say the least, since it strongly suggests that humanity is not necessarily predestined to an existence marked only by perpetual conflicts, wars, and genocides and the accompanying suffering and destruction they invariably produce.

In 1919, the Irish poet W. B. Yeats wrote a poem he titled "The Second Coming," in which he wrote the following lines:

> Turning and turning in the widening gyre
> The falcon cannot hear the falconer;
> Things fall apart; the centre cannot hold;
> Mere anarchy is loosed upon the world,
> The blood-dimmed tide is loosed, and everywhere
> The ceremony of innocence is drowned;
> The best lack all conviction, while the worst
> Are full of passionate intensity.[75]

In many important ways, Yeats's imagery provides a powerful allegory for the consequences of humanity failing to meet the challenges of climate change. The world he imagined in the poem is a harsh and violent one. As I have sought to illustrate in this book, the upcoming years and decades will see tremendous challenges, but we shouldn't forget that humanity has always confronted such challenges. For all of our faults as a species, we are remarkably versatile, adaptable, creative, and resilient. These are all qualities that will stand us in good stead. The trick is to find ways to resist the leaders, processes, and mentalities that have been shown to lead to intolerance, violence, and conflict, and to nurture and accelerate those that lead to cooperation, tolerance, and nonviolent solutions. Our future quite literally depends upon it.

NOTES

INTRODUCTION

1. Barack Obama, "Remarks by the President at U.N. Climate Change Summit," The White House, Office of the Press Secretary, September 24, 2014, accessed September 24, 2016, https://www.whitehouse.gov/the-press-office/2014/09/23/remarks-president-un-climate-change-summit.

2. John T. Abatzoglou and Park Williams, "Impact of Anthropogenic Climate Change on Wildfire across Western US Forests," *Proceedings of the National Academy of Sciences of the United States*, 2016.

3. Don E. Albrecht, "Population Brief, Trends in the Western U.S., the Changing West: A Regional Overview," Western Rural Development Center, 2008, accessed November 28, 2016, http://extension.oregonstate.edu/sites/default/files/for-employees/administrative-resources/diversity/The_Changing_West_2008_Regional_Overview.pdf.

4. The research literature on climate change in terms of articles and books is huge and growing daily. For readers interested in exploring the science of climate change, I would recommend a visit to the website for the Intergovernmental Panel on Climate Change. This organization is an international body of leading scientists who collect, organize, and publish the most up-to-date information on climate change research. Their publications are available for download at https://www.ipcc.ch/index.htm. For those interested in a less technical overview of the science behind climate change, I would recommend *The Thinking Person's Guide to Climate Change* by Robert Henson. In this book, Henson does an excellent job of summarizing in clear, accessible language the science, issues, and controversies surrounding climate change. Robert Henson,

The Thinking Person's Guide to Climate Change (Boston: The American Meteorological Society, 2014).

5. Kurt M. Campbell and Christine Parthemore, "National Security and Climate Change in Perspective," in *Climatic Cataclysm: The Foreign Policy and National Security Implications of Climate Change,* ed. Kurt M. Campbell (Washington, D.C.: Brookings Institution Press, 2008), 18.

I. MAKING SENSE OF CLIMATE CHANGE

1. Gavin Schmidt and Joshua Wolfe, *Climate Change: Picturing the Science* (New York: W. W. Norton and Company, 2009).

2. William James Burroughs, *Climate Change: A Multidisciplinary Approach,* 2nd ed. (New York: Cambridge University Press, 2007), 1.

3. E. Kirsten Peters, *The Whole Story of Climate: What Science Reveals About the Nature of Endless Change* (Amherst, N.Y.: Prometheus Books, 2012).

4. Dale Rice, "List of Record Low Temperatures Set Tuesday," *USA Today*, January 7, 2014, accessed August 11, 2014, http://www.usatoday.com/story/weather/2014/01/07/weather-polar-vortex-cold/4354945/.

5. Jon Erdman, "Lake Superior Still Has Ice despite Air Temperatures in the 80s on Memorial Day Weekend," *Weather.com*, May 27, 2014, accessed August 13, 2014, http://www.weather.com/news/lake-superior-ice-memorial-day-weekend-2014-20140526.

6. Justin Grieser, "Freezing Cold March Sets Records across Europe," *The Washington Post*, March 29, 2013, accessed October 23, 2014, http://www.washingtonpost.com/blogs/capital-weather-gang/wp/2013/03/29/cold-march-sets-records-across-europe/?print=1.

7. Jim Hoft, "Despite Coldest Winter in 100 Years—The Tolerant Left Wants Global Warming 'Deniers' Jailed," *The Gateway Pundit*, March 20, 2014, accessed August 13, 2014, http://www.thegatewaypundit.com/2014/03/despite-coldest-winter-in-100-years-the-tolerant-left-wants-global-warming-deniers-jailed/.

8. Samantha Motano, "The Louisiana Floods Are Devastating, and Climate Change Will Bring More Like Them," *Vox.com*, August 23, 2016, accessed September 9, 2016, http://www.vox.com/2016/8/18/12522036/louisiana-flood-climate-change-emergency-management.

9. Henry Fountain, "Scientists See Push from Climate Change in Louisiana Flooding," *The New York Times*, September 7, 2016, accessed September 9, 2016, http://www.nytimes.com/2016/09/08/science/global-warming-louisiana-flooding.html.

10. Eric Holthaus, "The Point of No Return: Climate Change Nightmares Are Already Here," *Rolling Stone*, August 5, 2016, accessed September 9, 2016, http://www.rollingstone.com/politics/news/the-point-of-no-return-climate-change-nightmares-are-already-here-20150805?utm_source=email.

11. See for example Greg Dotson and Erin Auel, "Trump Tests Climate Change Denial Against Public Opinion, Real-World Impacts," ThinkProgress.com, September 2, 2016, accessed September 9, 2016, https://thinkprogress.org/trump-climate-change-denial-public-opinion-impacts-3e096afd1264#.ad5ebeqap.

12. BrainyQuote.com, s.v. "Nicholas Stern," accessed September 24, 2016, http://www.brainyquote.com/quotes/quotes/n/nicholasst516728.html.

13. Juliane L. Fry, Hans-F Graf, Richard Grotjahn, Marilyn N. Raphael, Clive Saunders, and Richard Whitaker, *The Encyclopedia of Weather and Climate Change: A Complete Visual Guide* (Berkeley: University of California Press, 2010).

14. Burroughs, *Climate Change*.

15. J. Lean, "Evolution of the Sun's Spectral Irradiance since the Maunder Minimum," *Geophysics Research Letters* 27 (2000): 2425–28.

16. Brian Fagan, *The Long Summer: How Climate Changed Civilization* (New York: Basic Books, 2004).

17. Paul Reiter, "From Shakespeare to Defoe: Malaria in England in the Little Ice Age," *Emerging Infectious Diseases* 6, no. 1 (January-February 2000): 1–11.

18. Mark Maslin, "How the Ice Age Began," in *The Complete Ice Age: How Climate Change Shaped the World,* ed. Brian Fagan (New York: Thames and Hudson, 2009), chap. 3.

19. Tim Flannery, *The Weather Makers: How Man Is Changing the Climate and What It Means for Life on Earth* (New York: Grove Press, 2005).

20. Mark Maslin, "The Climatic Rollercoaster," in *The Complete Ice Age: How Climate Change Shaped the World,* chap. 4.

21. Fry et al., *The Encyclopedia of Weather*.

22. Doug Macdougal, *Frozen Earth: The Once and Future Story of Ice Ages* (Berkeley: University of California Press, 2004).

23. Brian Fagan, introduction to *The Complete Ice Age: How Climate Change Shaped the World*.

24. Susan Solomon, Dahe Qin, Martin Manning, Melinda Marquis, Kristen Averyt, Melinda M. B. Tignor, Henry LeRoy Miller Jr., and Zhenlin Chen, eds., *Climate Change 2007: The Physical Science Basis, Contribution of Working Group 1 to the Fourth Assessment Report of the Intergovernmental Panel on Climate Change* (New York: Cambridge University Press, 2007).

25. Brian Fagan, *Cro-Magnon: How the Ice Age Gave Birth to the First Modern Humans* (New York: Bloomsbury Press, 2010).

26. See for example Stanley H. Ambrose, "Did the Super-Eruption of Toba Cause a Human Population Bottleneck? Reply to Gathorne-Hardy and Harcourt-Smith," *Journal of Human Evolution* 45 (2003): 231–37.

27. See for example the World Meteorological Organization (WMO) and United Nations Environment Programme (UNEP) Intergovernmental Panel on Climate Change, http://www.ipcc.ch/index.htm.

28. See for example Climate Central, *Global Weirdness: Severe Storms, Deadly Heat Waves, Relentless Drought, Rising Seas, and the Weather of the Future* (New York: Vintage Books, 2012).

29. See for example Michael E. Mann and Lee R. Kump, *Dire Predictions: Understanding Climate Change. The Visual Guide to the Findings of the IPCC*, 2nd ed. (New York: DK Publishing, 2016).

30. Flannery, *The Weather Makers*.

31. Edward Wong, "Glut of Coal-Fired Plants Casts Doubts on China's Energy Priorities," *The New York Times*, November 11, 2015, accessed February 18, 2016, http://www.nytimes.com/2015/11/12/world/asia/china-coal-power-energy-policy.html.

32. Flannery, *The Weather Makers*.

33. Flannery, *The Weather Makers*.

34. Fry et al., *The Encyclopedia of Weather*.

35. Fry et al., *The Encyclopedia of Weather*.

36. Tom Vanderbilt, *Traffic: Why We Drive The Way We Do (and What It Says About Us)* (New York: Vintage Books, 2008).

37. Robert Barr, "China Surpasses US as Top Energy Consumer," *NBC News.com*, June 8, 2011, accessed September 26, 2014, http://www.nbcnews.com/id/43327793/ns/business-oil_and_energy/t/china-surpasses-us-top-energy-consumer/#.VCWBmyldWRp.

38. D. Laffoley and J. M. Baxter, eds., "Explaining Ocean Warming: Causes, Scale, Effects, and Consequences," *International Union for Conservation of Nature and Natural Resources*, September 2016, accessed September 14, 2016, https://portals.iucn.org/library/sites/library/files/documents/2016–046_0.pdf.

39. See for example Harold Hensel, "High Methane Levels on July 20, 2014, around Latitude 60 North," *Arctic News*, July 21, 2014, accessed July 29, 2014, http://arctic-news.blogspot.com/.

40. Henry Pollack, *A World Without Ice* (New York: Avery Books, 2009).

41. T. F. Stocker, D. Qin, G.-K. Plattner, M. Tignor, S. K. Allen, J. Boschung, A. Nauels, Y. Xia, V. Bex, and P. M. Midgley, eds., *Climate Change 2013: The Physical Science Basis. Contribution of Working Group I to the*

Fifth Assessment Report of the Intergovernmental Panel on Climate Change (New York: Cambridge University Press, 2013).

42. Brandon Miller, "2015 Is Warmest Year on Record, NOAA and NASA Say," *CNN.com*, January 20, 2016, accessed February 18, 2016, http://www.cnn.com/2016/01/20/us/noaa-2015-warmest-year/.

43. Ben Strauss, Claudia Tebaldi, and Remik Ziemlinski, "Surging Seas: Sea Level Rise, Storms & Global Warming's Threat to the US Coast: A Climate Central Report," *Climate Central*, March 14, 2012, accessed July 28, 2014, http://sealevel.climatecentral.org/.

44. Stocker et al., *Climate Change 2013*.

45. John Houghton, *Global Warming: The Complete Briefing*, 4th ed. (Cambridge: Cambridge University Press, 2009).

46. Burroughs, *Climate Change*.

47. Pollack, *A World Without Ice*.

48. Flannery, *The Weather Makers*.

49. Mark Jenkins. "Greenland: Ground Zero for Global Warming." *National Geographic*, June 2010.

50. Fry et al., *The Encyclopedia of Weather*.

51. Flannery, *The Weather Makers*.

52. Guy Raz, interview with Jane Ferrigno, "New Research Sheds Light on Antarctic Ice Melting," *All Things Considered*, NPR.org, February 28, 2010, accessed September 28, 2010, http://www.npr.org/templates/story/story.php?storyId=124178690.

53. Flannery, *The Weather Makers*.

54. Erik Conway, "Is Antarctica Melting?" NASA, January 12, 2010, accessed September 28, 2010, http://www.nasa.gov/topics/earth/features/20100108_Is_Antarctica_Melting.html.

55. Conway, "Is Antarctica Melting?"

56. James Hansen, Makiko Sato, Paul Hearty, Reto Ruedy, Maxwell Kelley, Valerie Masson-Delmotte, Gary Russell, George Tselioudis, Junji Cao, Eric Rignot, Isabella Velicogna, Blair Tormey, Bailey Donovan, Evgeniya Kandiano, Karina von Schuckmann, Pushker Kharecha, Allegra N. Legrande, Michael Bauer, and Kwok-Wai Lo, "Ice Melt, Sea Level Rise and Superstorms: Evidence from Paleoclimate Data, Climate Modeling, and Modern Observations That 2°C Global Warming Could Be Dangerous," *Journal of Atmospheric Chemistry and Physics* 16 (2016): 3761–812.

57. Kirstin Dow and Thomas E. Downing, *The Atlas of Climate Change: Mapping the World's Greatest Challenge* (Berkeley: University of California Press, 2007).

58. Maslin, "How the Ice Age Began."

59. While some have challenged the role of the Gulf Stream in helping create a mild climate in Europe, others continue to argue for the importance of the Gulf Stream in shaping European climate. See for example Richard Seager, "The Source of Europe's Mild Climate," *American Scientist*, 2006, accessed June 28, 2014, http://www.americanscientist.org/issues/feature/2006/4/the-source-of-europes-mild-climate/1. See also P. B. Rhines and S. Häkkinen, "Is the Oceanic Heat Transport in the North Atlantic Irrelevant to the Climate in Europe?" *ASOF Newsletter*, no. 1 (2003), accessed July 28, 2014, http://www.usclivar.org/sites/default/files/amoc/Rhines_Hakkinen_2003.pdf.

60. Maslin, "How the Ice Age Began."

61. Pollack, *A World Without Ice*.

62. Some suggest that the reason why Lake Agassiz spilled over was because it was hit by a comet, the impact of which caused the breach, but this theory is somewhat in dispute. See R. Firestone, A. West, J. Kennett, L. Becker, T. Bunch, Z. Revay, P. Schultz, T. Belgya, D. Kennett, J. Erlandson, O. Dickenson, A. Goodyear, R. Harris, G. Howard, J. Kloosterman, P. Lechler, P. Mayewski, J. Montgomery, R. Porede, T. Darrah, S. Hee, A. Smith, A. Stich, W. Topping, J. Wittke, and W. Wolbach, "Evidence for an Extraterrestrial Impact 12,900 Years Ago That Contributed to Megafaunal Extinctions and the Younger Dryas Cooling," *Proceedings of the National Academy of Sciences* 104, no. 41 (2007): 16016–21; Enrico de Lazaro, "Younger Dryas Climate Shift 12,900 Years Ago Linked to Asteroid or Comet Impact in Quebec," *Sci-News.com*, September 3, 2013, accessed August 1, 2014, http://www.sci-news.com/geology/science-younger-dryas-climate-shift-asteroid-comet-quebec-01351.html.

63. Burroughs, *Climate Change*; Mark Maslin, "How the Ice Age Began."

64. Gwynne Dyer, *Climate Wars: The Fight for Survival as the World Overheats* (Oxford: One World, 2010).

65. Rachel Warren, "The Role of Interactions in a World Implementing Adaptation and Mitigation Solutions to Climate Change," *Philosophical Transactions of the Royal Society A: Mathematical, Physical, and Engineering Sciences* 369 (2011): 217–41.

66. Joel K. Bourne Jr., *The End of Plenty: The Race to Feed a Crowded World* (New York: W. W. Norton & Company, 2015).

67. Thomas C. Peterson, Peter A. Stott, and Stephanie Herring, eds., "Explaining Extreme Events of 2011 from a Climate Perspective," *American Meteorological Society* (July 2012): 1041–67.

68. See for example Associated Press, "UN Says 2013 'Extreme' Weather Events Due to Human-Induced Climate Change," *Foxnews.com*, March 24, 2014, accessed August 3, 2014, http://www.foxnews.com/world/2014/03/24/un-says-2013-extreme-events-due-to-warming-earth-from-human-factors/.

69. For a good review of this argument see Bryan Walsh, "Climate Change Might Just Be Driving the Historic Cold Snap," *Time,* January 6, 2014, accessed August 13, 2014, http://science.time.com/2014/01/06/climate-change-driving-cold-weather/.

70. R. Nicholls, P. Wong, V. Burkett, J. Codignotto, J. Hay, R. McLean, S. Ragoonaden, and C. Woodroffe, "Coastal Systems and Low-Lying Areas," in *Climate Change 2007: Impacts, Adaptation and Vulnerability. Contribution of Working Group II to the Fourth Assessment Report of the Intergovernmental Panel on Climate Change*, ed. M. Parry, O. Canziani, J. Palutikof, P. van der Linden, and C. Hanson (Cambridge: Cambridge University Press, 2007), 315–56.

71. Anthony D. Barnosky, *Heatstroke: Nature in an Age of Global Warming* (Washington, D.C.: Island Press, 2009), 14–15.

72. Michela Pacifici, Piero Visconti, Stuart H. M. Butchart, James E. M. Watson, Francesca M. Cassola, and Carlo Rondinini, "Species Traits Influenced Their Response to Recent Climate Change," *Nature Climate Change* 7 (2017): 1–5, doi:10.1038/nclimate3223.

73. Pacifici et al., "Species Traits Influenced Their Response," 5.

74. For a good summary see Wolfgang Behringer, *A Cultural History of Climate*, trans. Patrick Camiller (Malden, Mass.: Polity Press, 2010).

75. See for example BBC, "Nature: Prehistoric Life," accessed August 14, 2014, http://www.bbc.co.uk/nature/history_of_the_earth; or Elizabeth Kolbert, *The Sixth Extinction: An Unnatural History* (New York: Henry Holt, 2014).

76. Kolbert, *The Sixth Extinction.*

77. Kolbert, *The Sixth Extinction.*

78. Terry L. Root, Jeff T. Price, Kimberly R. Hall, Stephen H. Schneider, Cynthia Rosenzweig, and J. Alan Pounds, "Fingerprints of Global Warming on Wild Animals and Plants," *Nature* 421 (2003): 57–60.

79. Root et al., "Fingerprints of Global Warming"; G. A. Rose, "On Distributional Responses of North Atlantic Fish to Climate Change," *ICES Journal of Marine Science* 62 (2005): 1360–74.

80. World Wildlife Fund, "Living Blue Planet Report," 2015, accessed February 18, 2016, http://www.wwf.or.jp/activities/upfiles/20150831LBPT.pdf.

81. Harald Welzer, *Climate Wars: Why People Will Be Killed in the 21st Century* (Malden, Mass.: Polity Press, 2012), 27.

82. Quoted in Kolbert, *The Sixth Extinction*, 268.

83. For a review of this literature see Behringer, *A Cultural History of Climate*.

84. Francis Jennings, *The Founders of America: From the Earliest Migrations to the Present* (New York: W. W. Norton and Company, 1994): Fagan,

The Long Summer; Ted Morgan, *Wilderness at Dawn: The Settling of the North American Continent* (New York: Touchstone Books, 1993).

85. Rita P. Wright, *The Ancient Indus: Urbanism, Economy, and Society* (Cambridge: Cambridge University Press, 2009); Gregory L. Possehl, *The Indus Civilization: A Contemporary Perspective* (Lanham, Md.: AltaMira Press, 2002).

86. Vasant Shinde, Shweta Sinha Deshpande, Toshiki Osada, and Takao Uno, "Basic Issues in Harappan Archaeology: Some Thoughts," *Ancient Asia* 1 (2006): 63–72, http://dx.doi.org/10.5334/aa.06107.

87. See Brian Fagan, *The Great Warming: Climate Change and the Rise and Fall of Civilizations* (New York: Bloomsbury Press, 2008).

88. Willam Rosen, *The Third Horseman: Climate Change and the Great Famine of the 14th Century* (New York: Viking Books, 2014).

89. James R. Lee, *Climate Change and Armed Conflict: Hot and Cold Wars* (New York: Routledge, 2009), 3.

2. ON THE ORIGINS OF VIOLENT CONFLICT

1. Christian Parenti, *Tropic of Chaos: Climate Change and the New Geography of Violence* (New York: Nation Books, 2011), 9.

2. Brian Fagan, *The Long Summer: How Climate Changed Civilization* (New York: Basic Books, 2004), 252.

3. Harald Welzer, *Climate Wars: Why People Will Be Killed in the 21st Century* (Malden, Mass.: Polity Press, 2012).

4. Robert W. Young and William Morgan, *The Navajo Language: A Grammar and Colloquial Dictionary* (Albuquerque: University of New Mexico Press, 1987).

5. David E. Stuart, *Anasazi America* (Albuquerque: University of New Mexico Press, 2000).

6. Joel K. Bourne Jr., *The End of Plenty: The Race to Feed a Crowded World* (New York: W.W. Norton and Company, 2015).

7. Karen Armstrong, *Fields of Blood: Religion and the History of Violence* (New York: Alfred A. Knopf, 2014), 12.

8. James F. Brooks, *Mesa of Sorrows: A History of the Awat'ovi Massacre* (New York: W. W. Norton and Company, 2016), 30.

9. Brooks, *Mesa of Sorrows*.

10. Christopher H. Guiterman, Thomas W. Swetnam, and Jeffrey S. Dean, "Eleventh-Century Shift in Timber Procurement Areas for the Great Houses of Chaco Canyon," *Proceedings of the National Academy of Sciences* 113, no. 5 (2016): 1186–90.

11. Christy G. Turner II and Jacqueline A. Turner, *Man Corn: Cannibalism and Violence in the Prehistoric American Southwest* (Salt Lake City: University of Utah Press, 1999).

12. Stuart, *Anasazi America*, 128.

13. Mark D. Varien, "Depopulation of the Northern San Juan Region: Historical Review and Archeological Context," in *Leaving Mesa Verde: Peril and Change in the Thirteenth-Century Southwest*, ed. Timothy A. Kohler, Mark D. Varien, and Aaron M. Wright (Tucson: University of Arizona Press, 2010); Aaron M. Wright, "The Climate of the Depopulation of the Northern Southwest," in *Leaving Mesa Verde.*

14. Kristin A. Kuckelman, "Catalysts of the Thirteenth-Century Depopulation of Sand Canyon Pueblo and the Central Mesa Verde Region," in *Leaving Mesa Verde*; Kristin A Kuckelman, "Thirteenth-Century Warfare in the Central Mesa Verde Region," in *Seeking the Center Place: Archeology and Ancient Communities in the Mesa Verde Region*, ed. Mark D. Varien and Richard H. Wilshusen (Salt Lake City: University of Utah Press, 2002).

15. William B. Carter, *Indian Alliances and the Spanish in the Southwest, 750–1750* (Norman: University of Oklahoma Press, 2009).

16. William Faulkner, *Requiem for a Nun* (New York: Random House, 1951).

17. Department of Defense, "National Security Implications of Climate-Related Risks and a Changing Climate" (July 23, 2015): 3, accessed July 17, 2016, http://archive.defense.gov/pubs/150724-congressional-report-on-national-implications-of-climate-change.pdf.

18. Department of Defense, "National Security Implications of Climate-Related Risks," 14.

19. Ian Rowlands, "The Security Challenges of Global Environmental Change," *Washington Quarterly* 14 (Winter 1991): 99.

20. United Nations High-Level Panel on Threats, Challenges and Change, *A More Secure World: Our Shared Responsibility* (New York: United Nations, 2005), 15.

21. See for example Alex Alvarez and Ronet Bachman, *Violence: The Enduring Problem*, 3rd ed. (Thousand Oaks, Calif.: Sage, 2017).

22. Herbert Blumer, "Collective Behavior," in *An Outline of the Principles of Sociology*, ed. R. E. Park (New York: Barnes and Noble, 1939).

23. Philip G. Zimbardo, "The Human Choice: Individuation, Reason, and Order vs. Deindividuation, Impulse, and Chaos," in *Nebraska Symposium on Motivation,* ed. W. J. Arnold and D. Levine (Lincoln: University of Nebraska Press, 1969), 237–307; Tom Postmes and Russell Spears, "Deindividuation and Antinormative Behavior: A Meta-Analysis," *Psychological Bulletin* 123 (1998):

238–59; James Waller, *Becoming Evil: How Ordinary People Commit Genocide and Mass Killing* (New York: Oxford University Press, 2002).

24. Werner Bergmann, "Pogroms," in *International Handbook of Violence Research*, vol. 1, ed. Wilhelm Heitmayer and John Hagan (Dordrecht: Kluwer Academic Publishers, 2003), 351–67.

25. Bradley Campbell, *The Geometry of Genocide: A Study in Pure Sociology* (Charlottesville: University of Virginia Press, 2015).

26. Jaskaran Kaur, "Twenty Years of Impunity: The November 1984 Pogroms of Sikhs in India," 2nd ed., *A Report by Ensaaf*, October 2006, accessed November 28, 2015, http://ensaaf-org.jklaw.net/publications/reports/20years/20years-2nd.pdf.

27. Kaur, "Twenty Years of Impunity."

28. In his work on ethnic and religious mobilization, political scientist Steven Wilkinson found that "religious mobilization around Hindu-Muslim issues and Sikh and Muslim identities is positively related to deaths," and suggests that much has to do with the way in which minorities are viewed in India and the unwillingness of government officials to intervene against antiminority violence. In fact, Wilkinson further suggests that political leaders have often capitalized upon and facilitated such violence. Steven I. Wilkinson, "Which Group Identities Lead to Most Violence? Evidence from India," in *Order, Conflict, and Violence*, ed. Stathis N. Kalyvas, Ian Shapiro, and Tarek Masoud (New York: Cambridge University Press, 2008), 293.

29. Shreeya Sinha and Mark Suppes, "Timeline of the Riot in Modi's Gujarat," *The New York Times*, August 19, 2015, accessed November 28, 2015, http://www.nytimes.com/interactive/2014/04/06/world/asia/modi-gujarat-riots-timeline.html.

30. Vicken Cheterian, *War and Peace in the Caucasus: Russia's Troubled Frontier* (London: Hurst, 2011).

31. Rene Lemarchand, *Burundi: Ethnic Conflict and Genocide* (New York: Cambridge University Press, 1996).

32. E. San Juan, *In the Wake of Terror: Class, Race, Nation, Ethnicity in the Postmodern World* (Lanham, Md.: Lexington Books, 2007).

33. Thomas Hobbes, "Chapter XIII: Of the Naturall Condition of Mankind," in *Leviathan*, 1651.

34. See for example Azar Gat, *War in Human Civilization* (New York: Oxford University Press, 2006).

35. Richard Wrangham and Dale Peterson, *Demonic Males: Apes and the Origins of Human Violence* (Boston: Mariner Books, 1996).

36. Lawrence H. Keeley, *War Before Civilization: The Myth of the Peaceful Savage* (New York: Oxford University Press, 1997).

37. For a discussion of the coevolution of society and war see Raymond C. Kelly, *Warless Societies and the Origin of War* (Ann Arbor: University of Michigan Press, 2003).

38. Carl Zimmer, "Agriculture Linked to DNA Changes in Ancient Europe," *The New York Times*, November 23, 2015, accessed November 26, 2015, http://www.nytimes.com/2015/11/24/science/agriculture-linked-to-dna-changes-in-ancient-europe.html.

39. Robert L. O'Connell, *Of Arms and Men: A History of War, Weapons, and Aggression* (New York: Oxford University Press, 1989).

40. Karen Armstrong, *Fields of Blood: Religion and the History of Violence* (New York: Alfred A. Knopf, 2014).

41. Stephen LeBlanc, *Constant Battles: The Myth of the Peaceful, Noble Savage* (New York: St. Martin's Press, 2003); Keeley, *War Before Civilization*.

42. Richard J. Evans, *The Third Reich in Power, 1933–1939* (New York: The Penguin Press, 2005); Ian Kershaw, *To Hell and Back: Europe 1914–1949* (New York: Viking Press, 2015).

43. There are a number of books that explicitly address this issue. See for example Gwynne Dyer, *Climate Wars: The Fight for Survival as the World Overheats* (Oxford: One World, 2010); Michael T. Klare, *Resource Wars: The New Landscape of Global Conflict* (New York: Henry Holt and Company, 2001); James R. Lee, *Climate Change and Armed Conflict: Hot and Cold Wars* (London: Routledge, 2009); Jeffrey Mazo, *Climate Conflict: How Global Warming Threatens Security and What to Do About It* (London: International Institute for Strategic Studies, 2010); Christian Parenti, *Tropic of Chaos: Climate Change and the New Geography of Violence* (New York: Nation Books, 2011).

44. Max Roser, "War and Peace after 1945," *Our World in Data.org*, 2016, accessed July 17, 2016, https://ourworldindata.org/war-and-peace-after-1945/; Jonathan Tepperman, *The Fix: How Nations Survive and Thrive in a World in Decline* (London: Bloomsbury Press, 2016).

45. Paul Collier, Anke Hoeffler, and Dominic Rohner, "Beyond Greed and Grievance: Feasibility and Civil War," *Oxford Economic Papers* 61, no. 1 (2009): 1–27.

46. Paul Collier, Anke Hoeffler, and Nicholas Sambanis, "The Collier-Hoeffler Model of Civil War Onset and the Case Study Project Research Design," in *Understanding Civil War: Evidence and Analysis. 1: Africa,* ed. Paul Collier and Nicholas Sambanis (Washington, D.C.: The World Bank, 2005).

47. Nicholas Sambanis, "Using Case Studies to Expand the Theory of Civil War," *CPR Working Papers*, Social Development Department, Environmentally and Socially Sustainable Development Network, Paper #5, May 2003,

accessed July 17, 2016, http://www.forecastingprinciples.com/files/pdf/Using_Case_Studies.pdf.

48. Martin Shaw, *War & Genocide* (Cambridge: Polity Press, 2003).

49. Lemkin was a lawyer and a Jew who had escaped the Nazi occupation of Poland and was profoundly concerned with the fate of his family and other victims of Nazi crimes. Donna-Lee Frieze, ed., *Totally Unofficial: The Auto-biography of Raphael Lemkin* (New Haven, Conn.: Yale University Press, 2013).

50. Lawrence J. LeBlanc, *The United States and the Genocide Convention* (Durham, N.C.: Duke University Press, 1991).

51. According to the United Nations Genocide Convention, genocide is legally defined as

> any of the following acts committed with intent to destroy, in whole or in part, a national, ethnical, racial or religious group, as such:
>
> a. killing members of the group;
> b. causing serious bodily or mental harm to members of the group;
> c. deliberately inflicting on the group conditions of life calculated to bring about its physical destruction in whole or in part;
> d. imposing measures intended to prevent births within the group;
> e. forcibly transferring children of the group to another group.

The complete text is available at the United Nations website at http://www.un.org/millenium/law/iv-1.htm.

52. R. Aron, *The Century of Total War* (New York: Doubleday, 1954); J. Black, *The Age of Total War, 1860–1945* (Westport, Conn.: Praeger Security International, 2006).

53. R. J. Rummel, *Death by Government* (New Brunswick, N.J.: Transaction Publishers, 1994).

54. Benjamin Valentino, *Final Solutions: Mass Killing and Genocide in the 20th Century* (Ithaca, N.Y.: Cornell University Press, 2004).

55. Helen Fein, *Genocide: A Sociological Perspective* (Thousand Oaks, Calif.: Sage, 1990).

56. See for example Gary Clayton Anderson, *Ethnic Cleansing and the Indian: The Crime That Should Haunt America* (Norman: University of Oklahoma Press, 2014); Alex Alvarez, *Native America and the Question of Genocide* (Lanham, Md.: Rowman & Littlefield, 2014).

57. See for example Christian Gerlach, *Extremely Violent Societies: Mass Violence in the Twentieth-Century World* (New York: Cambridge University Press, 2010).

58. See for example Alex Alvarez, "Genocide in the Context of War," in *The Palgrave Handbook of Criminology and War*, ed. Ross McGarry and Sandra Walklate (London: Palgrave Macmillan, 2016).

59. Taner Akçam, *A Shameful Act: The Armenian Genocide and the Question of Turkish Responsibility* (New York: Metropolitan Books, 2006); Taner Akçam, *The Young Turks' Crime Against Humanity: The Armenian Genocide and Ethnic Cleansing in the Ottoman Empire* (Princeton, N.J.: Princeton University Press, 2012); Donald Bloxham, *The Great Game of Genocide: Imperialism, Nationalism, and the Destruction of the Ottoman Armenians* (New York: Oxford University Press, 2005); Robert Melson, *Revolution and Genocide: On the Origins of the Armenian Genocide and the Holocaust* (Chicago: University of Chicago Press, 1992).

60. Stephen G. Fritz, *Ostkrieg: Hitler's War of Extermination in the East* (Lexington: University Press of Kentucky, 2011); Donald M. McKale, *Hitler's Shadow War: The Holocaust and World War II* (New York: Cooper Square Press, 2002); Alexander B. Rossino, *Hitler Strikes Poland: Blitzkrieg, Ideology, and Atrocity* (Lawrence: University Press of Kansas, 2003); Jeff Rutherford, *Combat and Genocide on the Eastern Front: The German Infantry's War, 1941–1944* (New York: Cambridge University Press, 2014).

61. Ben Kiernan, *The Pol Pot Regime: Race, Power, and Genocide in Cambodia under the Khmer Rouge, 1975–1979* (New Haven, Conn.: Yale University Press, 1996).

62. Alison des Forges, *Leave None to Tell the Story: Genocide in Rwanda* (New York: Human Rights Watch, 1999); Linda Melvern, *Conspiracy to Murder: The Rwandan Genocide* (London: Verso Books, 2004); Gerard Prunier, *Africa's World War: Congo, the Rwandan Genocide, and the Making of a Continental Catastrophe* (New York: Oxford University Press, 2009).

63. Richard J. Evans, *The Coming of the Third Reich* (New York: The Penguin Press, 2004), 172.

64. Götz Aly, *Why the Germans? Why the Jews? Envy, Race Hatred, and the Prehistory of the Holocaust* (New York: Metropolitan Books, 2011); Peter Fritsche, *Germans into Nazis* (Cambridge: Harvard University Press, 1998).

65. Karl A. Schleunes, *The Twisted Road to Auschwitz* (Urbana: University of Illinois Press, 1990); Saul Friedländer, *Nazi Germany and the Jews*, vol. 1, *The Years of Persecution, 1933–1939* (New York: HarperCollins Publishers, 1997).

66. Friedländer, *Nazi Germany and the Jews*.

67. Götz Aly and Susanne Heim, *Architects of Annihilation: Auschwitz and the Logic of Destruction* (Princeton, N.J.: Princeton University Press, 2002); Martin Winstone. *The Dark Heart of Hitler's Europe* (New York: I. B. Taurus, 2015).

68. Mark Roseman, *The Wannsee Conference and the Final Solution: A Reconsideration* (New York: Metropolitan Books, 2002).

69. Rossino, *Hitler Strikes Poland*.

70. Christopher R. Browning, *The Origins of the Final Solution: The Evolution of Nazi Jewish Policy, September 1939–March 1942* (Lincoln: University of Nebraska Press, 2004).

71. United States Holocaust Memorial Museum, *Historical Atlas of the Holocaust* (New York: Macmillan Publishing, 1996).

72. Chris Bellamy, *Absolute War: Soviet Russia and the Second World War* (New York: Alfred A. Knopf, 2007).

73. Steven Lehrer, *Wannsee House and the Holocaust* (Jefferson, N.C.: McFarland & Co., 2000); Roseman, *The Wannsee Conference and the Final Solution*.

74. Sarah K. Danielsson, "Creating Genocidal Space: Geographers and the Discourse of Annihilation, 1880–1933," *Space and Polity* 13, no. 1 (April 2009): 55–68.

75. Cited in Adam Tooze, *The Wages of Destruction: The Making and Breaking of the Nazi Economy* (New York: Penguin Books, 2006), 479.

76. Cited in Fritz, *Ostkrieg*, 62.

77. Fritz, *Ostkrieg*.

78. See for example Alex J. Kay, Jeff Rutherford, and David Stahel, eds., *Nazi Policy on the Eastern Front, 1941: Total War, Genocide, and Radicalization* (Rochester, N.Y.: University of Rochester Press, 2012).

79. Aaron T. Beck, *Prisoners of Hate: The Cognitive Basis of Anger, Hostility, and Violence* (New York: HarperCollins Publishers, 1999), 206.

80. Welzer, *Climate Wars*, 47.

81. See Jonathan Glover, *Humanity: A Moral History of the Twentieth Century* (New Haven, Conn.: Yale University Press, 1999).

82. B. Berkeley, "Ethnicity and Conflict in Africa: The Methods Behind the Madness," in *War Crimes: The Legacy of Nuremberg*, ed. B. Cooper (New York: TV Books, 1999); Gerard Prunier, *The Rwanda Crisis: History of a Genocide* (New York: Columbia University Press, 1995); David Rieff, *Slaughterhouse: Bosnia and the Failure of the West* (New York: Touchstone Books, 1995); Richard Rhodes, *Masters of Death: The SS-Einsatzgruppen and the Invention of the Holocaust* (New York: Alfred A. Knopf, 2002); Tom Segev, *Soldiers of Evil: The Commandants of the Nazi Concentration Camps* (New York: McGraw-Hill, 1987).

83. Glover, *Humanity*, 282.

84. This speech is available on YouTube in the original German with side-by-side translation.

85. Alex Hinton, *Why Did They Kill? Cambodia in the Shadow of Genocide* (Berkeley: University of California Press, 2005), 288.

86. Ervin Staub, *The Roots of Evil: The Origins of Genocide and Other Group Violence* (New York: Cambridge University Press, 1989).

87. Staub, *The Roots of Evil*, 16.

3. LINKING CLIMATE CHANGE
AND CONFLICT

1. Thomas Homer-Dixon, *Environment, Scarcity, and Violence* (Princeton, N.J.: Princeton University Press, 1999), 12.

2. John Medcof and John Roth, eds., *Approaches to Psychology* (Milton Keynes, UK: Open University Press, 1979), 206.

3. Abram De Swaan, *The Killing Compartments: The Mentality of Mass Murder* (New Haven, Conn.: Yale University Press, 2015), 48.

4. Tom Gjelten, *Sarajevo Daily: A City and Its Newspaper Under Siege* (New York: HarperCollins, 1995).

5. David Rieff, *Slaughterhouse: Bosnia and the Failure of the West* (New York: Touchstone Books, 1995), 105.

6. See for example Richard West, *Tito and the Rise and Fall of Yugoslavia* (New York: Carroll and Graf, 1994).

7. Andre Gerolymatos, *The Balkan Wars: Myth, Reality, and the Eternal Conflict* (Toronto: Stoddard Press, 2001); Ferdinand Schevill, *A History of the Balkans* (New York: Dorset Press, 1991).

8. Noel Malcolm, *Bosnia: A Short History* (New York: New York University Press, 1996), 174.

9. Malcolm, *Bosnia: A Short History*, 174.

10. Norman Cigar, *Genocide in Bosnia: The Policy of Ethnic Cleansing* (College Station: Texas A&M Press, 1995), 78.

11. Quoted in Cigar, *Genocide in Bosnia*, 78.

12. Vamik Volkan, *Blood Lines: From Ethnic Pride to Ethnic Terrorism* (New York: Farrar, Straus, and Giroux, 1997).

13. Timothy Judah, *The Serbs: History, Myth & the Destruction of Yugoslavia* (New Haven, Conn.: Yale University Press, 1997), 30.

14. Louis Sell, *Slobodan Milosevic and the Destruction of Yugoslavia* (Durham, N.C.: Duke University Press, 2002), 20.

15. Richard West, *Tito and the Rise and Fall of Yugoslavia* (New York: Carroll and Graf, 1994).

16. Michael A. Sells, *The Bridge Betrayed: Religion and Genocide in Bosnia* (Berkeley: University of California Press, 1996).

17. In Susan L. Woodward, *Balkan Tragedy: Chaos and Dissolution after the Cold War* (Washington, D.C.: The Brookings Institution, 1995), 37.

18. Laura Silber and Allan Little, *Yugoslavia: Death of a Nation* (New York: TV Books, 1995).

19. Woodward, *Balkan Tragedy*.

20. Silber and Little, *Yugoslavia: Death of a Nation*, 26.

21. Tone Bringa, *Being Muslim the Bosnian Way: Identity and Community in a Central Bosnian Village* (Princeton, N.J.: Princeton University Press, 1995), 3.

22. Bringa, *Being Muslim the Bosnian Way*.

23. Woodward, *Balkan Tragedy*.

24. Mark Danner, *Stripping Bare the Body: Politics Violence War* (New York: Nation Books, 2009).

25. Aleksa Djilas, "The Nation That Wasn't," in *The Black Book of Bosnia: The Consequences of Appeasement*, ed. Nader Mousavizadeh (New York: Basic Books, 1996), 24.

26. Dusko Doder and Louise Branson, *Milosevic: Portrait of a Tyrant* (New York: Free Press, 1999); Warren Zimmermann, *Origins of a Catastrophe* (New York: Times Books, 1999).

27. Julian Borger, *The Butcher's Trail: How the Search for Balkan War Criminals Became the World's Most Successful Manhunt* (New York: Other Press, 2016).

28. Lenard J. Cohen, *Serpent in the Bosom: The Rise and Fall of Slobodan Milošević* (Boulder, Colo: Westview, 2001); Doder and Branson, *Milosevic: Portrait of a Tyrant*; Adam LeBor, *Milosevic: A Biography* (New Haven, Conn.: Yale University Press, 2004); Sell, *Slobodan Milosevic and the Destruction of Yugoslavia*.

29. LeBor, *Milosevic: A Biography*, 121.

30. Woodward, *Balkan Tragedy*.

31. Roger Cohen, *Hearts Grown Brutal: Sagas of Sarajevo* (New York: Random House, 1998).

32. For a more extensive discussion see Alex Alvarez, *Governments, Citizens, and Genocide: A Comparative and Interdisciplinary Approach* (Bloomington: Indiana University Press, 2001).

33. Silber and Little, *Yugoslavia: Death of a Nation*.

34. Report of the Commission of Experts Pursuant to United Nations Security Council resolution 780, May 21, 1994, http://www.un.org/ga/search/view_doc.asp?symbol=S/1994/674.

35. Jan Willem Honig and Norbert Both, *Srebrenica: Record of a War Crime* (New York: Penguin Books, 1996); David Rohde, *Endgame: The Betrayal and Fall of Srebrenica: Europe's Worst Massacre Since World War II*

(New York: Farrar, Straus, and Giroux, 1997); Eric Stover and Gilles Peress, *The Graves: Srebrenica and Vukovar* (Berlin: Scalo Press, 1998).

36. Patrick Ball, Ewa Tabeau, and Philip Verwimp, *The Bosnian Book of Dead: Assessment of the Database (Full Report)*. HICN Research Design Note 5, June 17, 2007, Households in Conflict Network, The Institute of Development Students at the University of Sussex.

37. Christian Parenti, *Tropic of Chaos: Climate Change and the New Geography of Violence* (New York: Nation Books, 2011), 7.

38. James R. Lee, *Climate Change and Armed Conflict: Hot and Cold Wars* (New York: Routledge, 2009), 3.

39. Jeffrey Mazo, *Climate Conflict: How Global Warming Threatens Security and What to Do About It* (New York: Routledge, 2010).

40. Thomas M. Nichols, *Eve of Destruction: The Coming Age of Preventive War* (Philadelphia: University of Pennsylvania Press, 2008).

41. Alex Alvarez and Ronet Bachman, *Violence: The Enduring Problem*, 3rd ed. (Thousand Oaks, Calif.: Sage, 2017).

42. Gianfranco Poggi, "The Nation-State," in *Comparative Politics*, ed. Daniele Caramani (New York: Oxford University Press, 2008).

43. Anthony Giddens, *The Nation-State and Violence*, vol. 2, *A Contemporary Critique of Historical Materialism* (Berkeley: University of California Press, 1987), 120.

44. Patrick Stewart, *Weak Links: Fragile States, Global Threats, and International Security* (Oxford: Oxford University Press, 2011).

45. Michael Burton, Richard Gunther, and John Higley, "Introduction: Elite Transformations and Democratic Regimes," in *Elites and Democratic Consolidation in Latin America and Southern Europe*, ed. John Higley and Richard Gunther (Cambridge: Cambridge University Press, 1992), 10.

46. Daniel Rotberg, "Failed States in a World of Terror," *Foreign Affairs*, July/August 2002, accessed July 28, 2016, https://www.foreignaffairs.com/articles/2002-07-01/failed-states-world-terror.

47. Fund for Peace, Fragile States Index 2015, accessed September 21, 2016, http://fsi.fundforpeace.org/rankings-2015.

48. Jack Goldstone, Ted Robert Gurr, Barbara Harff, Marc A. Ley, Monty G. Marshall, Robert H. Bates, David L. Epstein, Colin H. Kahl, Pamela T. Surko, John C. Ulfelder, and Alan N. Unger, "State Failure Task Force Report: Phase III Findings," September 30, 2000, first presented at Science Applications International Corporation, McLean, Virginia, accessed November 24, 2016, https://www.hks.harvard.edu/fs/pnorris/Acrobat/stm103%20articles/StateFailureReport.pdf.

49. Goldstone et al., "State Failure Task Force Report: Phase III Findings."

50. Lance H. Gunderson, "Ecological Resilience—In Theory and Application," *Annual Review of Ecology and Systematics* 31 (November 2000): 425–39.

51. Joy J. Burnham, "Contemporary Fears of Children and Adolescents: Coping and Resiliency in the 21st Century," *Journal of Counseling and Development* 87, no. 1(Winter 2009): 28–35; Sam Goldstein and Robert B. Brooks, eds., *Handbook of Resilience in Children*, 2nd ed. (New York: Springer Publishing, 2012); Annalakshmi Narayanan and Lucy R. Betts, "Bullying Behaviors and Victimization Experiences Among Adolescent Students: The Role of Resilience," *Journal of Genetic Psychology* 175, no. 2 (2014): 136–46.

52. For a review, see W. Neil Adger, "Social and Ecological Resilience: Are They Related?" *Progress in Human Geography* 24, no. 3 (2000): 347–64.

53. Mazo, *Climate Conflict*; Adger, "Social and Ecological Resilience."

54. Philippe Le Billon, *Fuelling War: Natural Resources and Armed Conflict* (New York: Routledge, 2005).

55. Daron Acemoglu and James A. Robinson, *Why Nations Fail: The Origins of Power, Prosperity, and Poverty* (New York: Crown Business, 2012).

56. Val Percival and Thomas Homer-Dixon, "Environmental Scarcity and Violent Conflict: The Case of South Africa," *Journal of Peace Research* 35, no. 3 (May 1998): 281.

57. Robert I. Rotberg, "Failed States in a World of Terror," *Foreign Affairs*, July 1, 2002, accessed February 22, 2016, https://www.foreignaffairs.com/articles/2002-07-01/failed-states-world-terror.

58. Robert H. Bates, *When Things Fell Apart: State Failure in Late-Century Africa* (New York: Cambridge University Press, 2008).

59. Bates, *When Things Fell Apart*.

60. Bates, *When Things Fell Apart*, 133.

61. Thomas F. Homer-Dixon, *Environment, Scarcity, and Violence* (Princeton, N.J.: Princeton University Press, 1999).

62. Homer-Dixon, *Environment, Scarcity, and Violence*.

63. Michael T. Klare, *Resource Wars: The New Landscape of Global Conflict* (New York: Owl Books, 2001).

64. Thomas C. Hayes, "Confrontation in the Gulf: The Oilfield Lying Below the Iraq-Kuwait Dispute," *The New York Times*, September 3, 1990, accessed July 28, 2016, http://www.nytimes.com/1990/09/03/world/confrontation-in-the-gulf-the-oilfield-lying-below-the-iraq-kuwait-dispute.html.

65. See Alison Des Forges, *Leave None to Tell the Story: Genocide in Rwanda* (New York: Human Rights Watch, 1999); Linda Melvern, *Conspiracy to Murder: The Rwandan Genocide* (London: Verso, 2004).

66. Karol Boudreaux, "Land Conflict and Genocide in Rwanda," *The Electronic Journal of Sustainable Development* 1, no. 3 (2009).

67. Lee Ann Fujii, *Killing Neighbors: Webs of Violence in Rwanda* (Ithaca, N.Y.: Cornell University Press, 2009); Jean Hatzfeld, *Machete Season: The Killers in Rwanda Speak* (New York: Picador, 2003).

68. Martin Meredith, *The Fate of Africa: A History of the Continent Since Independence* (New York: Public Affairs Press, 2011).

69. William Maley, *What Is a Refugee?* (London: Hurst & Company, 2016).

70. Jason K. Stearns, *Dancing in the Glory of Monsters: The Collapse of the Congo and the Great War of Africa* (New York: Public Affairs, 2011).

71. See the International Rescue Committee, which conducted a number of mortality surveys between 2000 and 2007, http://www.irc.org.

72. Nicholas Kristof, "The Weapon of Rape," *The New York Times* Op-Ed, June 15, 2008, http://www.nytimes.com/2008/06/15/opinion/15kristof.html?_r=2&oref=slogin.

73. Jeffrey Gettleman, "The World's Worst War," *The New York Times*, December 15, 2012, accessed October 1, 2016, http://www.nytimes.com/2012/12/16/sunday-review/congos-never-ending-war.html.

74. See for example Gérard Prunier, *From Genocide to Continental War: The "Congolese" Conflict and the Crisis of Contemporary Africa* (London: Hurst and Company, 2011).

75. Stearns, *Dancing in the Glory of Monsters*.

76. Global Witness, "Natural Resource Exploitation and Human Rights in the Democratic Republic of the Congo, 1993–2003," *Global Witness Briefing Paper*, December 2009, accessed October 1, 2016, https://www.globalwitness.org/sites/default/files/pdfs/drc_exploitation_and_human_rights_abuses_93_03_en.pdf.

77. Michael T. Klare, *The Race for What's Left: The Global Scramble for the World's Last Resources* (New York: Picador, 2012).

78. See for example Paul McMahon, *Feeding Frenzy: Land Grabs, Price Spikes, and the World Food Crisis* (Vancouver: Greystone Books, 2014).

79. McMahon, *Feeding Frenzy*.

80. Luke Patey, *The New Kings of Crude: China, India, and the Global Struggle for Oil in Sudan and South Sudan* (London: Hurst & Company, 2014).

81. See for example, "Why Is the South China Sea Contentious?" *BBC News*, July 12, 2016, accessed October 1, 2016, http://www.bbc.com/news/world-asia-pacific-13748349.

82. Michael T. Klare, *Resource Wars: The New Landscape of Global Conflict* (New York: Owl Books, 2001), 25.

83. Klare, *The Race for What's Left*, 214–15.

84. Scott Straus, *Making and Unmaking Nations: War, Leadership, and Genocide in Modern Africa* (Ithaca, N.Y.: Cornell University Press, 2015).

85. Icek Ajzen and Martin Fishbein, *Understanding Attitudes and Predicting Social Behavior* (Englewood Cliffs, N.J.: Prentice Hall, 1980).

86. Jack Katz, *Seductions of Crime: Moral and Sensual Attractions in Doing Evil* (New York: Basic Books, 1988).

87. Alex Alvarez, *Governments, Citizens, and Genocide: A Comparative and Interdisciplinary Approach* (Bloomington: Indiana University Press, 2001); J. Waller, *Becoming Evil: How Ordinary People Commit Genocide and Mass Killing* (New York: Oxford University Press, 2002).

88. Michael Ignatieff, *Blood and Belonging: Journeys into the New Nationalism* (New York: The Noonday Press, 1993), 3.

89. Bruce Porter, *War and the Rise of the State: The Military Foundations of Modern Politics* (New York: Free Press, 1994), 123.

90. Vamik Volkan, *Blood Lines: From Ethnic Pride to Ethnic Terrorism* (New York: Farrar, Straus, and Giroux, 1997), 23.

91. Göran Therborn, *The Ideology of Power and the Power of Ideology* (London: NLB, 1980), 9.

92. Alvarez, *Governments, Citizens, and Genocide*; Alex Alvarez, *Genocidal Crimes* (New York: Routledge, 2010).

93. Peter Singer, *The Expanding Circle: Ethics and Sociobiology* (New York: Straus & Giroux, 1981).

94. Helen Fein, *Genocide: A Sociological Perspective* (London: Sage Publications, 1993).

95. Albert Bandura, "Moral Disengagement in the Perpetration of Inhumanities," *Personality and Social Psychology* Review 3 (1999): 193–209.

96. Alvarez, *Governments, Citizens, and Genocide*.

97. Alvarez, *Governments, Citizens, and Genocide*; Herbert C. Kelman, and V. Lee Hamilton, *Crimes of Obedience: Toward a Social Psychology of Authority and Responsibility* (New Haven, Conn.: Yale University Press, 1989); James Waller, *Becoming Evil: How Ordinary People Commit Genocide and Mass Killing*, 2nd ed. (New York: Oxford University Press, 2007).

98. Volkan, *Blood Lines*.

99. H. Hirsch, *Genocide and the Politics of Memory: Studying Death to Preserve Life* (Chapel Hill: University of North Carolina Press, 1995), 26.

100. Erving Staub, *The Roots of Evil: The Origins of Genocide and Other Group Violence* (New York: Cambridge University Press, 1989), 14.

101. Ervin Staub, "The Psychology of Bystanders, Perpetrators, and Heroic Helpers," in *Understanding Genocide: The Social Psychology of the Holocaust*, ed. L. Newman and R. Erber (New York: Oxford University Press, 2002).

102. Michael T. Costelloe, Ted Chiricos, Jiří Buriánek, Marc Gertz, and Daniel Maier-Katkin, "The Social Correlates of Punitiveness Toward Criminals: A Comparison of the Czech Republic and Florida," *The Justice System*

Journal 23, no. 2 (2002): 191–220; Michael T. Costelloe, Ted Chiricos, and Marc Gertz, "Punitive Attitudes Toward Criminals: Exploring the Relevance of Crime Salience and Economic Insecurity," *Punishment and Society* 11, no. 1 (2007): 25–29; Devon Johnson, "Anger about Crime and Support for Punitive Criminal Justice Policies," *Punishment and Society* 11, no. 1 (2007): 51–66.

103. Randall Collins, *Violence: A Micro-Sociological Theory* (Princeton, N.J.: Princeton University Press, 2008).

104. David Garland, *The Culture of Control: Crime and Social Order in Contemporary Society* (Chicago: University of Chicago Press, 2001).

105. J. Carroll, *Constantine's Sword: The Church and the Jews* (Boston: Houghton Mifflin Company, 2001); D. Prager and J. Telushkin, *Why the Jews: The Reason for Antisemitism* (New York: Touchstone Books, 1983); J. Weiss, *Ideology of Death: Why the Holocaust Happened in Germany* (Chicago: Elephant Paperbacks, 1996).

106. R. Rhodes, *Masters of Death: The SS-Einsatzgruppen and the Invention of the Holocaust* (New York: Alfred A. Knopf, 2002), 35.

107. See for example Curt R. Bartol and Anne M. Bartol, *Criminal Behavior: A Psychosocial Approach*, 7th ed. (Upper Saddle River, N.J.: Pearson/Prentice Hall, 2005).

108. See for example Gregor Aisch, Adam Pearce, and Bryant Rousseau, "How Far Is Europe Swinging to the Right," *The New York Times*, July 5, 2016, accessed August 25, 2016, http://www.nytimes.com/interactive/2016/05/22/world/europe/europe-right-wing-austria-hungary.html.

109. Manuel Funke, Moritz Schularick, and Christoph Trebesch, "Going to Extremes: Politics after Financial Crises, 1870–2014," *European Economic Review* 88 (2016): 227–60.

110. Oli Brown, *Migration and Climate Change* (Geneva: International Organization for Migration, 2008); Koko Warner et al., "In Search of Shelter: Mapping the Effects of Climate Change on Human Migration and Displacement," United Nations University, CARE, and CIESIN-Columbia University, 2009, http://ciesin.columbia.edu/documents/clim-migr-report-june09_final.pdf.

111. Georg Rusche and Otto Kirchheimer, *Punishment and Social Structure* (New York: Russell and Russell, 1968).

112. See for example Michael Mann, *The Dark Side of Democracy: Explaining Ethnic Cleansing* (New York: Cambridge University Press, 2005).

113. Jerrold M. Post, *Leaders and Their Followers in a Dangerous World: The Psychology of Political Behavior* (Ithaca, N.Y.: Cornell University Press, 2004).

4. WATER, VIOLENT CONFLICT,
AND GENOCIDE

1. Serageldin was the World Bank's vice president for environmental affairs as well as serving as the chair of the World Water Commission. Quoted in Marq De Villiers, *Water: The Fate of Our Most Precious Resource* (Boston: Houghton Mifflin, 2000), 13.

2. Brian Fagan, *Elixir: A History of Water and Humankind* (New York: Bloomsbury Press, 2011), xiii.

3. James Salzman, *Drinking Water: A History* (New York: Overlook Duckworth, 2012), 47.

4. Peter Beaumont, "Mohammed Bouazizi: The Dutiful Son Whose Death Changed Tunisia's Fate," *The Guardian*, January 20, 2011, accessed January 8, 2016, http://www.theguardian.com/world/2011/jan/20/tunisian-fruit-seller-mohammed-bouazizi.

5. Kareem Fahim, "Slap to a Man's Pride Set Off Tumult in Tunisia," *The New York Times*, January 21, 2011, accessed January 8, 2016, http://www.nytimes.com/2011/01/22/world/africa/22sidi.html.

6. Fahim, "Slap to a Man's Pride."

7. Yasmine Ryan, "The Tragic Life of a Street Vendor: Al Jazeera Travels to the Birthplace of Tunisia's Uprising and Speaks to Mohammed Bouazizi's Family," *Al Jazeera*, January 20, 2011, accessed June 20, 2016, http://www.aljazeera.com/indepth/features/2011/01/201111684242518839.html.

8. Rania Abouzeid, "Bouazizi: The Man Who Set Himself and Tunisia on Fire," *Time Magazine,* January 21, 2011, accessed June 20, 2016, http://content.time.com/time/magazine/article/0,9171,2044723,00.html.

9. Robert F. Worth, "How a Single Match Can Ignite a Revolution," *The New York Times*, January 21, 2011, accessed January 6, 2016, http://www.nytimes.com/2011/01/23/weekinreview/23worth.html.

10. John Carlo Gil Sadian, "The Arab Spring—One Year Later: The CenSEI Report Analyzes How 2011's Clamor for Democratic Reform Met 2012's Need to Sustain Its Momentum," *The CenSEI Report*, February 13, 2012, accessed January 8, 2016, http://www.scribd.com/doc/90470593/The-CenSEI-Report-Vol-2-No-6-February-13-19-2012#outer_page_23.

11. "Islam's Old Schism, Sunnis v Shias, Here and There: Sectarian Rivalry, Reverberating in the Region, Is Making Many Muslims Reflect," *The Economist*, June 29, 2013, accessed January 10, 2016, http://www.economist.com/news/middle-east-and-africa/21580162-sectarian-rivalry-reverberating-region-making-many-muslims.

12. Julia Zorthian, "Who's Fighting Who in Syria," *Time*, October 7, 2015, accessed October 1, 2016, http://time.com/4059856/syria-civil-war-explainer/.

13. Lucy Rodgers, David Gritten, James Offer, and Patrick Asare, "Syria: The Story of the Conflict," *BBC News*, March 11, 2016, accessed June 20, 2016, http://www.bbc.com/news/world-middle-east-26116868.

14. Peter H. Gleick, "Water, Drought, Climate Change, and Conflict in Syria," *American Meteorological Society* 6 (July 2014): 331–40.

15. Arab Center for the Studies of Arid Zones and Dry Lands, "Drought Vulnerability in the Arab Region: Case Study—Drought in Syria, Ten Years of Scarce Water (2000–2010)," April 2011, accessed January 10, 2016, http://www.unisdr.org/files/23905_droughtsyriasmall.pdf.

16. Arab Center, "Drought Vulnerability in the Arab Region."

17. Mahmoud Solh, "Tackling the Drought in Syria," *Nature Middle East, Emerging Science in the Arab World*, September 27, 2010, accessed January 10, 2016, http://www.natureasia.com/en/nmiddleeast/article/10.1038/nmiddleeast.2010.206.

18. Suzanne Saleeby, "Sowing the Seeds of Dissent: Economic Grievances and the Syrian Social Contract's Unraveling," *Socialist Viewpoint* 12, no. 2 (March/April 2012), accessed January 10, 2016, http://www.socialistviewpoint.org/marapr_12/marapr_12_33.html.

19. Eric Hoffer, *The True Believer: Thoughts on the Nature of Mass Movements* (New York: Harper Perennial, 1951), 3.

20. Benjamin Franklin, Founders' Quotes.com, accessed March 27, 2017, http://foundersquotes.com/founding-fathers-quote/when-the-wells-dry-we-know-the-worth-of-water/.

21. Thomas A. Miller, *Modern Surgical Care: Physiologic Foundations and Clinical Applications,* 3rd ed. (New York: Informa Healthcare, 2006).

22. U.S. Geological Survey, "The USGS Water Science School," accessed May 27, 2016, http://water.usgs.gov/edu/propertyyou.html.

23. Anne C. Grandjean, "Water Requirements, Impinging Factors, and Recommended Intakes" (World Health Organization, 2004).

24. Duncan Ryan, *The Sumerians: History's First Recorded Civilization* (Amazon Digital Services: Kindle Book, 2016).

25. Fagan, *Elixir*; Cynthia Barnett, *Rain: A Natural and Cultural History* (New York: Crown Books, 2015).

26. Fagan, *Elixir.*

27. Steven Solomon, *Water: The Epic Struggle for Wealth, Power, and Civilization* (New York: Harper, 2010).

28. Solomon, *Water: The Epic Struggle.*

29. Thomas H. Maugh, "Migration of Monsoons Created, Then Killed Harappan Civilization," *Los Angeles Times,* May 28, 2012, accessed January 19, 2016, http://articles.latimes.com/2012/may/28/science/la-sci-sn-indus-harappan-20120528.

30. UN Water.org, "Water Cooperation: Facts and Figures," http://www. unwater.org/water-cooperation-2013/water-cooperation/facts-and-figures/en/.

31. UN Water.org, "Water Cooperation."

32. World Health Organization (WHO), Media Centre, "Drinking Water," Fact Sheet No. 391, June 2015, accessed January 6, 2015, http://www.who.int/ mediacentre/factsheets/fs391/en/.

33. WHO, "Drinking Water."

34. Salzman, *Drinking Water*.

35. Juliane Fry, Hans-F Graf, Richard Grotjahn, Marilyn N. Raphael, Clive Saunders, and Richard Whitaker, *The Encyclopedia of Weather and Climate Change: A Complete Visual Guide* (Berkeley: University of California Press, 2010).

36. WHO, "Drinking Water."

37. Salzman, *Drinking Water*. 19.

38. Maggie Black and Jannet King, *The Atlas of Water: Mapping the World's Most Critical Resource* (Berkeley: University of California Press, 2009).

39. Advances in preserving and transporting food supplies also played a significant role.

40. Joel K. Bourne Jr., *The End of Plenty: The Race to Feed a Crowded World* (New York: W. W. Norton & Company, 2015).

41. Natural Resources Management and Environment Department, "Water and Food Security," United Nations Food and Agriculture Organization, accessed August 29, 2016, http://www.fao.org/docrep/x0262e/x0262e01.htm.

42. D. S. Battisti and R. L. Naylor, "Historical Warnings of Food Insecurity with Unprecedented Seasonal Heat," *Science* 323, no. 5911 (January 9, 2009): 240–44; Stephen P. Long, Elizabeth A. Ainsworth, Andrew D. B. Leakey, Josef Nösberger, and Donald R. Ort, "Food for Thought: Lower-Than-Expected Crop Yield Stimulation with Rising CO_2 Concentrations," *Science* 312, no. 5782 (June 30, 2006): 1918–21.

43. Black and King, *The Atlas of Water*.

44. Vivek Singh, "Parched Land. Farmer Suicides. Forced Migration: Drought Is Crippling Rural India," *The World Post*, June 30, 2016, accessed July 2, 2016, http://www.huffingtonpost.com/entry/india-drought-bundelkhand_us_5769a229e4b0c0252e778b9a.

45. Sonya Fatah, "Suicide Rate Growing as Debt Cripples India's Farms," *Toronto Star*, March 24, 2008, accessed July 8, 2016, https://www.thestar.com/ news/world/2008/03/24/suicide_rate_growing_as_debt_cripples_indias_farms. html.

46. Michael Kimmelman, "Climate Change Is Threatening to Push a Crowded Capital Toward a Breaking Point," *The New York Times*, February

17, 2017, accessed February 17, 2017, https://www.nytimes.com/interactive/2017/02/17/world/americas/mexico-city-sinking.html.

47. Black and King, *The Atlas of Water*.

48. Christopher Boucek, "Yemen: Avoiding a Downward Spiral," Carnegie Papers, no. 102 (Washington, D.C.: Carnegie Endowment for International Peace, September 2009).

49. Robert E. Kopp, Andrew C. Kemp, Klaus Bittermann, Benjamin P. Horton, Jeffrey P. Donnelly, W. Roland Gehreis, Carling C. Hay, Jerry X. Mitrovica, Eric D. Morrow, and Stefan Rahmsdorf, "Temperature-Driven Global Sea-Level Variability in the Common Era," *Proceedings of the National Academy of Sciences*, February 22, 2016, accessed July 2, 2016, http://www.pnas.org/content/113/11/E1434.full.pdf; Matthias Mengel, Anders Levermann, Katja Frieler, Alexander Robinson, and Ben Marzeion, "Future Sea Level Rise Constrained by Observations and Long-Term Commitment," *Proceedings of the National Academy of Sciences* 113, no. 10 (March 8, 2016), accessed July 2, 2016, http://www.pnas.org/content/113/10/2597.full.pdf.

50. Elizabeth Kolbert, "Greenland Is Melting," *The New Yorker*, October 24, 2016, accessed November 1, 2016, http://www.newyorker.com/magazine/2016/10/24/greenland-is-melting.

51. James Hansen, Makiko Sato, Paul Hearty, Reto Ruedy, Maxwell Kelley, Valerie Masson-Delmotte, Gary Russell, George Tselioudis, Junji Cao, Eric Rignot, Isabella Velicogna, Blair Tormey, Bailey Donovan, Evgeniya Kandiano, Karina von Schuckmann, Pushker Kharechia, Allegra N. Legrande, Michael Bauer, and Kwok-Wai Lo, "Ice Melt, Sea Level Rise and Superstorms: Evidence from Paleoclimate Data, Climate Modeling, and Modern Observations That 2°C Global Warming Could Be Dangerous," *Atmospheric Chemistry and Physics* 16 (March 22, 2016): 3761–812, accessed July 2, 2016, http://www.atmos-chem-phys.net/16/3761/2016/acp-16-3761-2016.pdf.

52. Justin Gillis, "Flooding of Coast, Caused by Global Warming, Has Already Begun," *The New York Times*, September 3, 2016, accessed September 5, 2016, http://www.nytimes.com/2016/09/04/science/flooding-of-coast-caused-by-global-warming-has-already-begun.html.

53. Hansen et al., "Ice Melt, Sea Level Rise and Superstorms."

54. Adam Sobel, *Storm Surge: Hurricane Sandy, Our Changing Climate, and Extreme Weather of the Past and Future* (New York: Harper Wave, 2014).

55. Kirsten Dow and Thomas E. Downing, *The Atlas of Climate Change: Mapping the World's Greatest Challenge* (Berkeley: University of California Press, 2007).

56. Barbara Neumann, Athanasios T. Vafeidis, Juliane Zimmermann, and Robert J. Nicholls, "Future Coastal Population Growth and Exposure to Sea-

Level Rise and Coastal Flooding—A Global Assessment," *PLOS One* 10, no. 3 (March 11, 2015).

57. R. Nicholls, P. Wong, V. Burkett, J. Codignotto, J. Hay, R. McLean, S. Ragoonaden, and C. Woodroffe, "Coastal Systems and Low-Lying Areas," in *Climate Change 2007: Impacts, Adaptation and Vulnerability. Contribution of Working Group II to the Fourth Assessment Report of the Intergovernmental Panel on Climate Change,* ed. M. Parry, O. Canziani, J. Palutikof, P. van der Linden, and C. Hanson (Cambridge: Cambridge University Press, 2007), 315–56.

58. J. Ericson, C. Vorosmarty, S. Dingman, L. Ward, and M. Meybeck, "Effective Sea-Level Rise and Deltas: Causes of Change and Human Dimension Implications," *Global Planet Change* 50 (2006): 63–82; C. Woodroffe, R. Nicholls, Y. Saito, Z. Chen, and S. Goodbred, "Landscape Variability and the Response of Asian Megadeltas to Environmental Change," in *Global Change and Integrated Coastal Management: The Asia-Pacific Region,* ed. N. Harvey (New York: Springer, 2006), 277–314; A. Sánchez-Arcilla, J. Jiménez, and H. Valdemoro, "A Note on the Vulnerability of Deltaic Coasts. Application to the Ebro Delta," in *Managing Coastal Vulnerability: An Integrated Approach,* ed. L. McFadden, R. Nicholls, and E. Penning-Rowsell (Amsterdam: Elsevier Science, 2006); A. Sánchez-Arcilla, J. Jiménez, H. Valdemoro, and V. Gracia, "Implications of Climatic Change on Spanish Mediterranean Low-Lying Coasts: The Ebro Delta Case," *Journal of Coastal Research* 24, no. 2 (2007): 306–16.

59. Anastasia Romanou, "Sea Changes," in *Climate Change: Picturing the Science,* ed. Gavin Schmidt and Joshua Wolfe (New York: W.W. Norton & Company, 2009), chap. 3.

60. Ericson et al., "Effective Sea-Level Rise."

61. Joanna I. Lewis, "China," in *Climate Change and National Security: A Country-Level Analysis,* ed. Daniel Moran (Washington, D.C.: Georgetown University Press, 2011), chap. 2.

62. George Friedman, *The Next 100 Years: A Forecast for the 21st Century* (New York: Anchor Books, 2009).

63. Corydon Ireland, "Rising Seas, Imperiled Cities: Coastal Regions Must Prepare, and Boston Is a Case Study," *Harvard University Center for the Environment* 3, no. 2 (Fall/Winter 2011/2012), accessed October 16, 2015, http://issuu.com/huce/docs/newsletter2011-final?e=3862434/2959944; William V. Sweet and Joseph Park, "From the Extreme to the Mean: Acceleration and Tipping Points of Coastal Inundation from Sea Level Rise," *Earth's Future,* December 18, 2014, accessed September 5, 2016, http://onlinelibrary.wiley.com/doi/10.1002/2014EF000272/full.

64. Justin Gillis, "Flooding of Coast, Caused by Global Warming, Has Already Begun," *The New York Times*, September 3, 2016, accessed September 5, 2016, http://www.nytimes.com/2016/09/04/science/flooding-of-coast-caused-by-global-warming-has-already-begun.html.

65. Nicholls et al., "Coastal Systems and Low-Lying Areas."

66. B. Lynn Ingram and Frances Malamud-Roam, *The West Without Water: What Past Floods, Droughts, and Other Climatic Clues Tell Us about Tomorrow* (Berkeley: University of California Press, 2013).

67. T. V. Paul, "India," in *Climate Change and National Security: A Country-Level Analysis*, ed. Daniel Moran (Washington, D.C.: Georgetown University Press, 2011), chap. 6.

68. Paul, "India," 75.

69. Christian Parenti, *Tropic of Chaos: Climate Change and the New Geography of Violence* (New York: Nation Books, 2011), 152.

70. Parenti, *Tropic of Chaos*.

71. Nida Najar, "Violence Erupts in Southern India over Order to Share Water," *The New York Times*, September 12, 2016, accessed September 13, 2016, http://www.nytimes.com/2016/09/13/world/asia/violence-erupts-in-southern-india-over-order-to-share-water.html.

72. Michael T. Klare, *Resource Wars: The New Landscape of Global Conflict* (New York: Metropolitan/Owl Books, 2001).

73. Michael B. Oren, *Six Days of War: June 1967 and the Making of the Modern Middle East* (New York: Oxford University Press, 2002).

74. Cited in Fred Pearce, *When the Rivers Run Dry: Water—The Defining Crisis of the Twenty-First Century* (Boston: Beacon Press, 2006).

75. Pearce, *When the Rivers Run Dry*; see also Chris McGreal, "Deadly Thirst," *The Guardian*, January 12, 2004, accessed July 14, 2016, https://www.theguardian.com/environment/2004/jan/13/water.israel.

76. Pearce, *When the Rivers Run Dry*.

77. Dawn Elizabeth Rehm and Farzano Noshab, "Pakistan," *Asian Development Outlook 2013: Asia's Energy Challenge* (Asian Development Bank, 2013), accessed August 27, 2016, http://www.adb.org/sites/default/files/publication/30205/ado2013-pakistan.pdf.

78. Aamir Saeed, "Rapid Groundwater Depletion Threatens Pakistan Food Security—Officials," *Reuters*, June 10, 2015, accessed August 27, 2016, http://www.reuters.com/article/pakistan-water-food-idUSL5N0YV3HQ20150610.

79. Pearce, *When the Rivers Run Dry*.

80. Pearce, *When the Rivers Run Dry*.

81. Palash Ghosh, "What Are India and Pakistan Really Fighting About?" *International Business Times*, December 27, 2013, accessed August 27, 2016,

http://www.ibtimes.com/what-are-india-pakistan-really-fighting-about-1520856.

82. Ghosh, "What Are India and Pakistan Really Fighting About?"

83. Population Division, *World Population Prospects: The 2015 Revision*, Department of Economic and Social Affairs, United Nations Secretariat, accessed August 29, 2016, https://esa.un.org/unpd/wpp/publications/files/key_findings_wpp_2015.pdf.

84. Daniel Markey, "Pakistan," in *Climate Change and National Security: A Country-Level Analysis*, ed. Daniel Moran (Washington, D.C.: Georgetown University Press, 2011).

85. Bourne Jr., *The End of Plenty*.

86. Bourne Jr., *The End of Plenty*.

87. Bourne Jr., *The End of Plenty*.

88. Marshall B. Burke, Edward Miguel, Shankar Satyanath, John A. Dykema, and David B. Lobell, "Warming Increases the Risk of Civil War in Africa," *Proceedings of the National Academy of Sciences* 6, no. 49 (December 8, 2009): 20670–74.

89. The World Bank, *World Development Report 2008: Agriculture for Development* (Washington, D.C.: The World Bank, 2007).

90. Brahma Chellaney, *Water, Peace, and War: Confronting the Global Water Crisis* (Lanham, Md.: Rowman & Littlefield, 2015), 5.

91. James R. Lee, *Climate Change and Armed Conflict: Hot and Cold Wars* (New York: Routledge, 2009); Jared Diamond, *Collapse: How Societies Choose to Fail or Succeed* (New York: Penguin Books, 2006).

92. Linda Melvern, *Conspiracy to Murder: The Rwandan Genocide* (London: Verso Books, 2004); Mahmood Mamdani, *When Victims Become Killers: Colonialism, Nativism, and the Genocide in Rwanda* (Princeton, N.J.: Princeton University Press, 2001); Deborah Mayerson, *On the Path to Genocide: Armenia and Rwanda Reexamined* (New York: Berghahn Books, 2014).

93. Mark Lynas, *Six Degrees: Our Future on a Hotter Planet* (Washington, D.C.: National Geographic, 2008).

94. Lynas, *Six Degrees*.

95. Specifically, they referred to it as Bilad al-Sudan or the Land of the Blacks. See Mahmood Mamdani, *Saviors and Survivors: Darfur, Politics, and the War on Terror* (New York: Pantheon Books, 2009).

96. Andrew S. Natsios, *Sudan, South Sudan, and Darfur: What Everyone Needs to Know* (New York: Oxford University Press, 2012).

97. Julie Flint and Alex De Waal, *Darfur: A New History of a Long War* (London: Zed Books, 2008).

98. Natsios, *Sudan, South Sudan, and Darfur*, 26.

99. Gérard Prunier, *Darfur: A 21st Century Genocide*, 3rd ed. (Ithaca, N.Y.: Cornell University Press, 2008).

100. Richard Cockett, *Sudan: Darfur and the Failure of an African State* (New Haven, Conn.: Yale University Press, 2010).

101. Jerry Fowler, "Evolution of Conflict and Genocide in Sudan: A Historical Survey," in *Darfur: Genocide Before Our Eyes*, ed. Joyce Apsel (New York: Institute for the Study of Genocide, 2005); Eric Reeves, "Darfur: Genocide Before Our Eyes," in *Darfur: Genocide Before Our Eyes*.

102. Cockett, *Sudan: Darfur and the Failure of an African State*; Natsios, *Sudan, South Sudan, and Darfur*.

103. Flint and De Waal, *Darfur: A New History*.

104. Abdullah El-Tom, "The Black Book of Sudan: Imbalance of Power and Wealth in Sudan," *Journal of African National Affairs* 1, no. 2 (2003): 25–35, accessed September 1, 2016, https://web.archive.org/web/20080516091953/http://www.ossrea.net/publications/newsletter/oct02/article9.htm.

105. Flint and De Waal, *Darfur: A New History*.

106. El-Tom, "The Black Book of Sudan."

107. M. W. Daly, *Darfur's Sorrow: A History of Destruction and Genocide* (New York: Cambridge University Press, 2007).

108. Prunier, *Darfur: A 21st Century Genocide*.

109. Prunier, *Darfur: A 21st Century Genocide*.

110. John Hagan and Winona Rymond-Richmond, *Darfur and the Crime of Genocide* (Cambridge: Cambridge University Press, 2009).

111. Quoted in Prunier, *Darfur: A 21st Century Genocide*, 100.

112. Prunier, *Darfur: A 21st Century Genocide*.

113. Fowler, "Evolution of Conflict and Genocide in Sudan."

114. Alex de Waal, *Famine That Kills: Darfur, Sudan* (New York: Oxford University Press, 2005).

115. J. Andrew Plowman, *Climate Change & Conflict Prevention* (Washington, D.C.: National Intelligence University, 2014).

116. Natsios, *Sudan, South Sudan, and Darfur*.

117. Natsios, *Sudan, South Sudan, and Darfur*, 118.

118. Plowman, *Climate Change & Conflict Prevention*.

119. Mamdani, *Saviors and Survivors*; Hagan and Rymond-Richmond, *Darfur and the Crime of Genocide*.

120. Eric Reeves, "Genocide by Attrition: Agony in Darfur," *Dissent* 52, no. 1 (2005): 21–25.

5. FORCED DISPLACEMENT AND BORDERS
IN A WARMING WORLD

1. Harald Welzer, *Climate Wars: Why People Will Be Killed in the 21st Century,* trans. Patrick Camiller (Malden, Mass.: Polity Press, 2012), 5.

2. Michael Werz and Laura Conley, "Climate Change, Migration, and Conflict: Addressing Complex Crisis Scenarios in the 21st Century," Center for American Progress, Heinrich Böll Stiftung (January 2012): 3–4, accessed August 5, 2016, https://www.americanprogress.org/issues/security/report/2012/01/03/10857/climate-change-migration-and-conflict/.

3. Eric Holthaus, "The Point of No Return: Climate Change Nightmares Are Already Here," *Rolling Stone*, August 5, 2015, accessed September 9, 2016, http://www.rollingstone.com/politics/news/the-point-of-no-return-climate-change-nightmares-are-already-here-20150805?utm_source=email.

4. Peter Andreas, *Border Games: Policing the U.S.-Mexico Divide,* 2nd ed. (Ithaca, N.Y.: Cornell University Press, 2009), 141.

5. The poem is titled *The New Colossus,* http://www.libertystatepark.com/emma.htm.

6. Karl Jacoby, "How the Border Threat Became Standard Political Trope," *Dallas News*, October 14, 2016, accessed October 21, 2016, http://www.dallasnews.com/opinion/commentary/2016/10/14/border-threat-became-standard-political-trope.

7. Peter Holley, "Trump Proposes a Border Wall. But There Already Is One, and It Gets Climbed Over," *The Washington Post*, April 2, 2016, accessed September 6, 2016, https://www.washingtonpost.com/news/morning-mix/wp/2016/04/02/shocking-video-shows-suspected-drug-smugglers-easily-crossing-u-s-mexico-border/.

8. Reece Jones, *Violent Borders: Refugees and the Right to Move* (London: Verso Books, 2016).

9. Anti-Defamation League, "Border Disputes: Armed Vigilantes in Arizona," 2003, accessed September 19, 2016, http://www.adl.org/assets/pdf/combating-hate/Border-Disputes-Armed-Vigilantes-in-Arizona.pdf.

10. Luis Alberto Urrea, *The Devil's Highway: A True Story* (New York: Back Bay Books, 2004).

11. Shuaizhang Feng, Alan B. Krueger, and Michael Oppenheimer, "Linkages Among Climate Change, Crop Yields and Mexico-US Cross-Border Migration," *Proceedings of the National Academy of Sciences of the United States of America* 107, no. 32 (August 10, 2010): 14257–62.

12. Richard Grant, *God's Middle Finger: Into the Lawless Heart of the Sierra Madre* (New York: Simon & Schuster, 2008), 242.

13. As one report summarizes, "Increasing gang violence, entrenched poverty and climate change are all factors that collude and collide, pushing more and more people out of their homes." Lauren Markham, "Drought and Climate Change Are Forcing Young Guatemalans to Flee to the U.S.," *The World Post*, February 16, 2017, accessed February 17, 2017, http://www. huffingtonpost.com/entry/climate-change-coffee-guatemala_us_ 589dd223e4b094a129ea4ea2?oupiku6nbdtcmzpvi.

14. Jeffrey S. Passel and D'Vera Cohn, "Unauthorized Immigrant Population Stable for Half a Decade," Pew Research Center, July 22, 2015, accessed September 6, 2016, http://www.pewresearch.org/fact-tank/2015/07/22/ unauthorized-immigrant-population-stable-for-half-a-decade/.

15. Faye Bowers, "U.S.-Mexican Border as a Terror Risk," *The Christian Science Monitor*, March 22, 2005, accessed September 20, 2016, http://www. csmonitor.com/2005/0322/p01s01-uspo.html; John Tirman, "Immigration and Insecurity: Post-9/11 Fear in the United States," *SSRC.org*, July 28, 2006, accessed September 20, 2016, http://borderbattles.ssrc.org/Tirman/printable. html.

16. Ioan Grillo, *El Narco: Inside Mexico's Criminal Insurgency* (New York: Bloomsbury Press, 2012).

17. J. Levin and G. Rabrenovic, *Why We Hate* (Amherst, N.Y.: Prometheus, 2004).

18. An important note about terminology is in order here. One important distinction is often made between migrants and those who are forcibly displaced. The term "migrant" is often used to refer to someone who voluntarily chooses to relocate in an attempt to find a better life, while those who are forcibly displaced have fled to escape a conflict or persecution. The category of forcibly displaced can include refugees who cross international borders, while internally displaced populations, or IDPs, refer to those who have moved but stay within the same nation. Some refugees may also be categorized as asylum seekers since they are attempting to find legal sanctuary in a new host country. Distinguishing between these categories is inherently difficult, yet one that has profound legal implications. Undocumented immigrants are often deported since it is assumed they are motivated by economic reasons and can therefore be returned without risk, while refugees have legal protections arising out of their political status as someone fleeing a war zone or some form of persecution. They are entitled to stay in the host nation while asylum proceedings are carried out. In some ways, such a distinction is problematic since one can question how voluntary is their choice when migrants risk a dangerous journey that is dictated by extreme poverty and economic need. Furthermore, the use of the term "illegal immigrant" is highly problematic for many people. Some prefer to use the terms "undocumented," "unlawfully present," "without stat-

us," or "noncitizens"; yet even these terms have their shortcomings and detractors. See for example Amanda Sakuma, "What's in a Name? Migrant vs. Refugee vs. Illegal Immigrant," *MSNBC.com*, May 19, 2016, accessed September 17, 2016, http://www.msnbc.com/msnbc/whats-name-migrant-vs-refugee-vs-illegal-immigrant.

19. Carlo Angerer and Alastair Jamieson, "71 Dead Refugees Found in Truck on Austrian Highway: Officials," *NBC News*, August 28, 2015, accessed November 2, 2016, http://www.nbcnews.com/storyline/europes-border-crisis/71-dead-refugees-found-truck-austria-highway-officials-n417536.

20. BBC, "Migrant Crisis: Migration to Europe Explained in Seven Charts," *BBCnews.com*, March 4, 2016, accessed September 17, 2016, http://www.bbc.com/news/world-europe-34131911.

21. Jones, *Violent Borders*.

22. IOM, "Mediterranean Migrant Arrivals in 2016: 204,311; Deaths 2,443," International Organization for Migration, May 31, 2016, accessed September 17, 2016, http://www.iom.int/news/mediterranean-migrant-arrivals-2016-204311-deaths-2443.

23. T. Brian and F. Laczko, *Fatal Journeys: Tracking Lives Lost During Migration* (Geneva: International Organization for Migration, 2014).

24. Matthew Carr, *Fortress Europe: Inside the War Against Immigration* (London: Hurst & Company, 2015).

25. Wolfgang Bauer, *Crossing the Sea with Syrians on the Exodus to Europe* (Los Angeles: And Other Stories Publishers, 2016).

26. Rick Noack, "Germany Welcomed More Than 1 Million Refugees in 2015. Now the Country Is Searching for Its Soul," *The Washington Post*, May 4, 2016, accessed September 19, 2016, https://www.washingtonpost.com/news/worldviews/wp/2016/05/04/germany-welcomed-more-than-1-million-refugees-in-2015-now-the-country-is-searching-for-its-soul/.

27. Bruce Katz, Luise Noring, and Nantke Garrelts, *Cities and Refugees—The German Experience* (Washington, D.C.: The Brookings Institution, September 2016).

28. Rick Noack, "Leaked Document Says 2,000 Men Allegedly Assaulted 1,200 German Women on New Year's Eve." *The Washington Post*, July 11, 2016, accessed September 19, 2016, https://www.washingtonpost.com/news/worldviews/wp/2016/07/10/leaked-document-says-2000-men-allegedly-assaulted-1200-german-women-on-new-years-eve/.

29. Katrin Elger, Ansbert Knelp, and Merlind Theile, "Survey Shows Alarming Lack of Integration in Germany," *Spiegel Online*, January 26, 2009, accessed September 20, 2016, http://www.spiegel.de/international/germany/immigration-survey-shows-alarming-lack-of-integration-in-germany-a-603588.html.

30. Heribert Adam, "Xenophobia, Asylum Seekers, and Immigration Policies in Germany," *Nationalism and Ethnic Politics* 21, no. 4 (2015): 446–64.

31. See for example Heribert Adam and Kogila Moodley, *Imagined Liberation: Xenophobia, Citizenship, and Identity in South Africa, Germany, and Canada* (Philadelphia: Temple University Press, 2015).

32. Robert D. Putnam, "*E Pluribus Unum*: Diversity and Community in the Twenty-First Century," *Scandinavian Political Studies* 30, no. 2 (2007): 137–74.

33. See for example Michelle Lynn Kahn, "The Cologne Sexual Assaults in Historical Perspective," *Notches*, January 19, 2016, accessed September 21, 2016, http://notchesblog.com/2016/01/19/the-cologne-sexual-assaults-in-historical-perspective/; Lauren Nelson, "'Islamic Rape of Europe' Magazine Cover Highlights Racism Behind Anti-Immigrant Sentiment," *Friendly Atheist*, February 19, 2016, accessed September 16, 2016, http://www.patheos.com/blogs/friendlyatheist/2016/02/19/islamic-rape-of-europe-magazine-cover-highlights-racism-behind-anti-immigrant-sentiment/.

34. Jones, *Violent Borders*.

35. Der Spiegel Staff, "Frustrations Grow over German Response to Terror," *Der Spiegel Online*, August 12, 2016, accessed September 19, 2016, http://www.spiegel.de/international/germany/frustrations-grow-over-german-response-to-terror-a-1107413.html.

36. See for example David Zucchino, "I've Become a Racist: Migrant Wave Unleashes Danish Tensions over Identity," *The New York Times*, September 5, 2016, accessed September 19, 2016, http://www.nytimes.com/2016/09/06/world/europe/denmark-migrants-refugees-racism.html; Aamna Mohdin, "Europe's Most Refugee-Friendly Country Is Growing Weary," *The World Post*, September 7, 2016, accessed September 19, 2016, http://www.huffingtonpost.com/entry/sweden-growing-weary-of-refugees_us_57d05741e4b06a74c9f22347.

37. The New York Times, "Europe's Rising Far Right: A Guide to the Most Prominent Parties," *The New York Times*, June 13, 2016, accessed September 19, 2016, http://www.nytimes.com/interactive/2016/world/europe/europe-far-right-political-parties-listy.html.

38. Caroline Copley, "German Government Fears Xenophobia Will Do Economic Harm: Violent Acts by Rightist Supporters Rose by 43% in 2015," *The World Post*, September 21, 2016, accessed September 21, 2016, http://www.huffingtonpost.com/entry/german-government-fears-xenophobia-will-do-economic-harm_us_57e280e5e4b0e28b2b513287?section=; Marialena Perpiraki, "Greek Journalists Say Neo-Fascist Party Members Attacked Them During Anti-Refugee Protest," *The World Post*, September 21, 2016, accessed September 23, 2016, http://www.huffingtonpost.com/entry/golden-dawn-

journalist-greece-attack_us_57def2e6e4b04a1497b50815?section=; Jonathan Tepperman, *The Fix: How Nations Survive and Thrive in a World in Decline* (London: Bloomsbury, 2016).

39. Abby Haglage, "Hate Crime Victimization Statistics Show Rise in Anti-Hispanic Crime," *The Daily Beast,* February 20, 2014, accessed September 23, 2016, http://www.thedailybeast.com/articles/2014/02/20/hate-crime-victimization-statistics-show-rise-in-anti-hispanic-crime.html; Liam Stack, "American Muslims Under Attack," *The New York Times*, February 15, 2016, accessed September 23, 2016, http://www.nytimes.com/interactive/2015/12/22/us/Crimes-Against-Muslim-Americans.html; Griff Witte, "A Punch. A Death. And Now a Fear in Britain That a Surge in Hate Crimes Is Here to Stay," *The Washington Post*, September 25, 2016, accessed September 25, 2016, http://www.msn.com/en-us/news/world/a-punch-a-death-and-now-a-fear-in-britain-that-a-surge-in-hate-crimes-is-here-to-stay/ar-BBwv6Vk?li=BBnb7Kz.

40. Christian Parenti, *Tropic of Chaos: Climate Change and the New Geography of Violence* (New York: Nation Books, 2011), 214.

41. J. McAdam, ed. *Climate Change & Displacement: Multidisciplinary Perspectives* (Oxford: Hart Publishing, 2010).

42. IPCC First Assessment Report, "Policymaker Summary of Working Group II (Potential Impacts of Climate Change)," 1990, p. 103, para. 5.0.10, https://www.ipcc.ch/ipccreports/1992%20IPCC%20Supplement/IPCC_1990_and_1992_Assessments/English/ipcc_90_92_assessments_far_wg_II_spm.pdf.

43. P. M. Dolukhanov, "The Pleistocene-Holocene Transition in Northern Eurasia: Environmental Changes and Human Adaptations," *Quaternary International* 41–42 (1997): 181–91; B. V. Geel, J. Buurman, and H. T. Waterbolk, "Archaeological and Palaeoecological Indications of an Abrupt Climate Change in the Netherlands, and Evidence for Climatological Teleconnections around 2650 BP," *Journal of Quaternary Science* 11, no. 6 (1996): 451–60; Y. N. Gribchenko and E. I. Kurenkova, "The Main Stages and Natural Environmental Setting of Late Paleolithic Human Settlement in Eastern Europe," *Quaternary International* 41–42 (1997): 173–79; B. Huntley, "Climatic Change and Reconstruction," *Journal of Quaternary Research* 14, no. 6 (1999): 513–20; D. R. Yesner, "Human Dispersal into Interior Alaska: Antecedent Conditions, Mode of Colonization and Adaptations," *Quaternary Science Reviews* 20, no. 1–3 (2001): 315–27; P. D. Tyson, J. Lee-Thorp, K. Holmgren, and J. F. Thackeray, "Changing Gradients of Climate Change in Southern Africa during the Past Millennium: Implications for Population Movements," *Climatic Change* 52, no. 1 (2002): 129–35; Brian Fagan, *The Long Summer: How Climate Changed Civilization* (New York: Basic Books, 2004); B. Smit and Y. Cai, "Climate Change and Agriculture in China," *Global Environmental*

Change 6, no. 3 (1996): 205–14; Don Fixico, *The American Indian Mind in a Linear World* (New York: Routledge, 2003).

44. Oli Brown, *Migration and Climate Change* (Geneva: International Organization for Migration, 2008).

45. Christian Aid, "Human Tide: The Real Migration Crisis," *Christian Aid Report*, 2007, accessed September 20, 2016, https://www.christianaid.org.uk/Images/human-tide.pdf.

46. Oli Brown, "Migration and Climate Change," *IOM Migration Research Series* no. 31, 2008. https://www.iom.cz/files/Migration_and_Climate_Change_-_IOM_Migration_Research_Series_No_31.pdf.

47. See for example Christian Aid, "Human Tide"; or Stephen Castles and Mark J. Miller, *The Age of Migration: International Population Movements in the Modern World* (Basingstoke, U.K.; Palgrave Macmillan, 2003).

48. See for example Keith Lowe, *Savage Continent: Europe in the Aftermath of World War II* (New York: Penguin Books, 2012); Ian Buruma, *Year Zero: A History of 1945* (New York: Penguin Books, 2013).

49. For a review of theories of migration, see Stephen Castles, Hein De Haas, and Mark J. Miller, *The Age of Migration: International Population Movements in the Modern World*, 5th ed. (New York: The Guilford Press, 2014).

50. Clionadh Raleigh, Lisa Jordan, and Idean Salehyan, "Assessing the Impact of Climate Change on Migration and Conflict," *Social Development Department of the World Bank Group*, March 2008, accessed September 21, 2016, http://siteresources.worldbank.org/EXTSOCIALDEVELOPMENT/Resources/SDCCWorkingPaper_MigrationandConflict.pdf; Clionadh Raleigh and Lisa Jordan, "Climate Change and Migration: Emerging Patterns in the Developing World," in *Social Dimensions of Climate Change: Equity and Vulnerability in a Warming World*, ed. Robin Mearns and Andrew Norton (Washington, D.C.: The World Bank, 2010), 103–31, accessed September 23, 2016, https://openknowledge.worldbank.org/handle/10986/2689.

51. W. N. Adger, "Institutional Adaptation to Environmental Risk under the Transition in Vietnam," *Annals of the Association of American Geographers* 90 (2000): 738–58; Robin Mearns and Andrew Norton, "Equity and Vulnerability in a Warming World: Introduction and Overview," in *Social Dimensions of Climate Change* (Washington, D.C.: The World Bank, 2010), 1–44, accessed September 23, 2016, https://openknowledge.worldbank.org/handle/10986/2689.

52. Robert McLeman and Barry Smit, "Migration as an Adaptation to Climate Change," *Climatic Change* 76, no. 1–2 (2006): 31–53.

53. McLeman and Smit, "Migration as an Adaptation to Climate Change."

54. For a thorough discussion of climate-induced population displacement see Jane McAdam, ed., *Climate Change and Displacement: Multidisciplinary Perspectives* (Portland, Ore.: Hart Publishing, 2012).

55. Alexander Betts, *Survival Migration: Failed Governance and the Crisis of Displacement* (Ithaca, N.Y.: Cornell University Press, 2013).

56. Stephen Castles, "Environmental Change and Forced Migration: Making Sense of the Debate," *UNHCR Working Papers* 70 (2002): 1–14; W. B. Wood, "Ecomigration: Linkages Between Environmental Change and Migration," in *Global Migrants and Global Refugees: Problems and Solutions*, ed. A. R. Zolberg and P. M. Benda (New York: Berghahn Books, 2001), 42–61.

57. Walter Kälin, "Internal Displacement," in *The Oxford Handbook of Refugee and Forced Migration Studies,* ed. Elena Fiddian-Qasmiyeh, Gil Loescher, Katy Long, and Nando Sigona (New York: Oxford University Press, 2016), chap. 13.

58. In J. Andrew Plowman, *Climate Change, Conflict, & Prevention* (Washington, D.C.: National Intelligence University Press, 2014).

59. S. Henry, P. Boyle, and E. F. Lambin, "Modelling Inter-Provincial Migration in Burkina Faso, West Africa: The Role of Socio-Demographic and Environmental Factors," *Applied Geography* 23, no. 2–3 (2003): 115–36; Sabine Perch-Nielsen, Michèle B. Bättig, and Dieter Imboden, "Exploring the Link Between Climate Change and Migration," *Climatic Change* 91 (2008): 375–93.

60. Walter Kälin, "Conceptualizing Climate-Induced Displacement," in *Climate Change & Displacement*, ed. McAdam, 81–103.

61. The Office of the United Nations High Commissioner for Refugees (UNHCR), *The State of the World's Refugees: In Search of Solidarity* (Oxford: Oxford University Press, 2012).

62. Sarah Kenyon Lischer, "Conflict and Crisis Induced Displacement," in *The Oxford Handbook of Refugee and Forced Migration Studies,* ed. Katy Long Loescher and Nando Sigona (New York: Oxford University Press, 2016), 317–29.

63. S. Schmeidl, "Exploring the Causes of Forced Migration: A Pooled Time-Series Analysis, 1971–1990," *Social Science Quarterly* 78, no. 2 (1997): 284–308.

64. Helen Fein, "Accounting for Genocide after 1945: Theories and Some Findings," *International Journal on Group Rights* 1, no. 2 (1993): 79–106; Kurt Jonassohn, "Prevention without Prediction," *Holocaust and Genocide Studies* 7, no. 1 (1993): 1–13.

65. Paul Mojzes, *Balkan Genocides: Holocaust and Ethnic Cleansing in the Twentieth Century* (Lanham, Md.: Rowman & Littlefield, 2011); Laura Silber and Allan Little, *Yugoslavia: Death of a Nation* (New York: TV Books, 1996);

Chuck Sudetic, *Blood and Vengeance: One Family's Story of the War in Bosnia* (New York: W. W. Norton & Company, 1998).

66. Jérôme Tubiana, "Darfur: A Conflict for Land?" in *War in Darfur and the Search for Peace*, ed. Alex De Waal (Cambridge, Mass.: Harvard University Press, 2007), 69.

67. Myron Weiner, "Bad Neighbors, Bad Neighborhoods: An Inquiry into the Causes of Refugee Flows," *International Security* 21, no. 1 (Summer 1996): 5–42.

68. Idean Salehyan and Kristian Skrede Gleditsch, "Refugee Flows and the Spread of Civil War," *International Organization* 60, no. 2 (April 2006): 335–62.

69. Rafael Reuveny, "Climate Change–Induced Migration and Violent Conflict," *Political Geography* 26, no. 6 (2007): 656–73.

70. UNHCR, *The State of the World's Refugees*.

71. "An Interview with Author Todd Miller on Climate Change and Migration," *No More Deaths.org*, October 11, 2016, accessed October 26, 2016, http://nomoredeaths.org/root-causes-an-interview-with-author-todd-miller-on-climate-change-and-migration/.

72. Michel Agier, *Managing the Undesirables: Refugee Camps and Humanitarian Government* (Malden, Mass.: Polity Press, 2011).

73. Emma Haddad, *The Refugee in International Society: Between Sovereigns* (Cambridge: Cambridge University Press, 2008).

74. See for example Maria Joao Guia and Maartje van der Woude, *Social Control and Justice: Crimmigration in the Age of Fear* (The Hague, Netherlands: Eleven International Publishing, 2013); Juliet Stumpf, "The Crimmigration Crisis: Immigrants, Crime, and Sovereign Power," *American University Law Review* 56, no. 2 (2006): 367–419.

75. Krista Schlyer, *Continental Divide: Wildlife, People, and the Border Wall* (College Station: Texas A & M University Press, 2012), 4.

76. Matthew Carr, *Fortress Europe: Inside the War Against Immigration* (London: Hurst & Company, 2015).

77. Paul Ganster, with David E. Lorey, *The U.S.-Mexican Border Today: Conflict and Cooperation in Historical Perspective*, 3rd ed. (Lanham, Md.: Rowman & Littlefield, 2016), xvii.

78. Sandro Mezzadra and Brett Neilson, *Border as Method, or, the Multiplication of Labor* (Durham, N.C.: Duke University Press, 2013).

79. Jan Philipp Reemtsma, "Neighbourly Relations as a Resource for Violence," *Eurozine*, November 2, 2005, accessed November 2, 2016, http://www.eurozine.com/pdf/2005-11-02-reemtsma-en.pdf.

80. Richard L. Rubenstein, *The Age of Triage: A Chilling History of Genocide from the Irish Famine to Vietnam's Boat People* (Boston: Beacon Press, 1983), 1.

81. Quoted in Alex Laban Hinton, *Why Did They Kill? Cambodia in the Shadow of Genocide* (Berkeley: University of California Press, 2005), 19.

82. Welzer, *Climate Wars*, 181.

83. Mark Levene, *The Crisis of Genocide*, vol. 1, *Devastation: The European Rimlands 1912–1938*, vol. 2, *Annihilation: The European Rimlands 1939–1953* (Oxford: Oxford UP, 2014).

84. Timothy Snyder, *Bloodlands: Europe Between Hitler and Stalin* (New York: Basic Books, 2010).

85. Michel Agier, *Managing the Undesirables: Refugee Camps and Humanitarian Government*, trans. David Fernbach (London: Polity Press, 2011).

86. George Friedman, *The Next Decade: Where We've Been . . . and Where We're Going* (New York: Doubleday, 2011).

87. Anwar Ali, "Vulnerability of Bangladesh to Climate Change and Sea Level Rise through Tropical Cyclones and Storm Surges," *Water, Air, & Soil Pollution* 92 (1996): 171–79; Ali Riaz, "Bangladesh," in *Climate Change and National Security: A Country-Level Analysis*, ed. Daniel Moran (Washington, D.C.: Georgetown University Press, 2011).

88. C. Ninno, P. Dorosh, L. Smith, and D. Roy, *The 1998 Floods in Bangladesh: Disaster Impacts, Household Coping Strategies, and Response, Research Report 122* (Washington, D.C.: International Food Policy Research Institute, 2001).

89. Ali, "Vulnerability of Bangladesh."

90. Ali, "Vulnerability of Bangladesh."

91. AFP, "Thousands Homeless in Cyclone-Hit Bangladesh," *Daily Mail.com*, May 22, 2016, accessed September 19, 2016, http://www.dailymail.co.uk/wires/afp/article-3602066/Five-dead-thousands-flee-Cyclone-Roanu-nears-Bangladesh.html.

92. Riaz, "Bangladesh."

93. Shardul Agrawala, Tomoko Ota, Ahsan Uddin Ahmed, Joel Smith, and Maaretn van Aalst, "Development and Climate Change in Bangladesh: Focus on Coastal Flooding and the Sundarbans," *Environment Directorate, Development Cooperation Directorate, Working Party on Global and Structural Policies, Working Party on Development Cooperation and Environment, Organization for Economic Cooperation and Development*, 2003, accessed September 19, 2016, http://www.oecd.org/environment/cc/21055658.pdf.

94. Riaz, "Bangladesh," 112.

95. Dennis C. Blair, "Annual Threat Assessment of the US Intelligence Community," Testimony before the Senate Select Committee on Intelligence, February 2, 2010. This quote is cited in Parenti, *Tropic of Chaos*, 139.

96. R. Cruz, H. Harasawa, M. Lal, S. Wu, Y. Anokhin, B. Punsalmaa, Y. Honda, M. Jafari, C. Li, and N. Huu Ninh, "Asia," in *Climate Change 2007: Impacts, Adaptation and Vulnerability. Contribution of Working Group II to the Fourth Assessment Report of the Intergovernmental Panel on Climate Change*, ed. M. Parry, O. Canziani, J. Palutikof, P. van der Linden, and C. Hanson (New York: Cambridge University Press, 2007), 469–506.

97. Alex Perry, *Falling Off the Edge: Travels Through the Dark Heart of Globalization* (New York: Bloomsbury Press, 2008).

98. Abram De Swaan, *Killing Compartments: The Mentality of Mass Murder* (New Haven, Conn.: Yale University Press, 2015).

99. Reuveny, "Climate Change–Induced Migration."

100. Perry, *Falling Off the Edge*.

101. Human Rights Watch, *"Trigger Happy": Excessive Use of Force by Indian Troops at the Bangladesh Border* (New York: Human Rights Watch, 2010), http://www.hrw.org/sites/default/files/reports/bangladesh1210Web.pdf; PTI, "Indo-Bangla Border Fencing Work to Finish by 2017," *The Indian Express*, June 25, 2016, accessed October 11, 2016, http://indianexpress.com/article/india/india-news-india/indo-bangla-border-fencing-work-to-finish-by-2017-2875548/.

102. Bidisha Banerjee, "The Great Wall of India," *Slate.com*, December 20, 2010, accessed October 11, 2016, http://www.slate.com/articles/health_and_science/green_room/2010/12/the_great_wall_of_india.html.

103. Parenti, *Tropic of Chaos*.

6. PREVENTING CONFLICT AND BUILDING RESILIENCE

1. E. Kirsten Peters, *The Whole Story of Climate: What Science Reveals About the Nature of Endless Change* (Amherst, N.Y.: Prometheus Books, 2012), 258.

2. Naomi Klein, *This Changes Everything: Capitalism vs the Climate* (New York: Simon & Schuster, 2014), 442.

3. Justin Worland, "Global CO_2 Concentration Passes Threshold of 400 ppm—And That's Bad For the Climate," *Time,* October 24, 2016, accessed October 25, 2016, http://time.com/4542889/carbon-dioxide-400-ppm-global-warming/?xid=newsletter-brief.

4. Chris D'Angelo, "'Earth on the Edge': EU Agency Confirms 2016 as Hottest Year on Record," *The Huffington Post*, January 5, 2017, accessed January 6, 2017, http://www.huffingtonpost.com/entry/2016-hottest-year-copernicus_us_586e9ddfe4b099cdb0fc1943.

5. See for example Oliver Milman, "Planet at Its Hottest in 115,000 Years Thanks to Climate Change, Experts Say," *The Guardian*, October 4, 2016, accessed October 12, 2016, https://www.theguardian.com/environment/2016/oct/03/global-temperature-climate-change-highest-115000-years?cmp=oth_b-aplnews_d-1.

6. For an interesting discussion of these possibilities see Gernot Wagner and Martin L. Weitzman, *Climate Shock: The Economic Consequences of a Hotter Planet* (Princeton, N.J.: Princeton University Press, 2015).

7. See for example Ruth DeFries, *The Big Ratchet: How Humanity Thrives in the Face of Natural Crisis* (New York: Basic Books, 2014).

8. See for example Herbert Hirsch, *Anti-Genocide: Building an American Movement to Prevent Genocide* (Westport, Conn.: Praeger, 2002); Sam Totten, ed., *Impediments to the Prevention and Intervention of Genocide*, vol. 9 of *Genocide: A Critical Bibliographic Review* (New Brunswick, N.J.: Transaction Publishers, 2013); Scott Straus, *Fundamentals of Genocide and Mass Atrocity Prevention* (Washington, D.C.: United States Holocaust Museum, 2016); Neal Riemer, ed., *Protection Against Genocide: Mission Impossible?* (Westport, Conn.: Praeger, 2000); Gareth Evans, *The Responsibility to Protect: Ending Mass Atrocity Crimes Once and For All* (Washington, D.C.: Brookings Institution, 2008); David A. Hamburg, *Preventing Genocide: Practical Steps Toward Early Detection and Effective Action* (Boulder, Colo.: Paradigm Publishers, 2008); Madeleine Albright and William S. Cohen, *Preventing Genocide: A Blueprint for U.S. Policymakers* (Washington, D.C.: United States Holocaust Memorial Museum, 2008); James Waller, *Confronting Evil: Engaging Our Responsibility to Prevent Genocide* (New York: Oxford University Press, 2016).

9. Milton Leitenberg, "Deaths in Wars and Conflicts in the 20th Century," Cornell University Peace Studies Occasional Paper no. 29, 3rd ed., 2006, http://cissm.umd.edu/papers/files/deathswarsconflictsjune52006.pdf.

10. Rudy J. Rummel, *Death by Government* (New Brunswick, N.J.: Transaction Publishers, 1994), 9.

11. Leitenberg, "Deaths in Wars and Conflicts."

12. Rummel, *Death by Government*.

13. Eric Reeves, "Reckoning the Costs: How Many Have Died during Khartoum's Genocidal Counter-Insurgency in Darfur? What Has Been Left in the Wake of This Campaign?" *Sudan: Research, Analysis, and Advocacy*, March 8, 2016, accessed November 15, 2016, http://sudanreeves.org/2016/03/18/reckoning-the-costs-how-many-have-died-during-khartoums-genocidal-

counter-insurgency-in-darfur-what-has-been-left-in-the-wake-of-this-campaign/.

14. Icasualties.org, accessed January 14, 2014, http://icasualties.org/Iraq/index.aspx,.

15. Iraqbodycount.org, accessed January 4, 2014, http://www.iraqbodycount.org/.

16. Adapted from Hugo Slim, *Killing Civilians: Method, Madness, and Morality in War* (New York: Columbia University Press, 2008).

17. Daya Somasundaram and Sambasivamoorthy Sivayokan, "Rebuilding Community Resilience in a Post-War Context: Developing Insight and Recommendations—A Qualitative Study in Northern Sri Lanka," *International Journal of Mental Health Systems* 7, no. 3 (2013): 1–24; Bonnie L. Green, Matthew Friedman, Joop de Jong, Susan D. Solomon, Terence M. Keane, John A. Fairbank, Brigid Donelan, and Ellen Frey-Wouters, eds., *Trauma, Interventions in War and Peace: Prevention, Practice, and Policy* (New York: Kluwer Press, 2003); Bonnie L. Green, "Psychosocial Research in Traumatic Stress: An Update," *Journal of Trauma Stress* 7 (1994): 341–62.

18. Andrew Baum and India Fleming, "Implications of Psychological Research on Stress and Technological Accidents," *American Psychologist* 48, no. 6 (1993): 665–72; Ellen Gerrity and Brian Flynn, "Mental Health Consequences of Disasters," in *The Public Health Consequences of Disasters*, ed. E. Noji (New York: Oxford University Press, 1997), 101–21; Elizabeth M. Smith and Carol S. North, "Posttraumatic Stress Disorder in Natural Disasters and Technological Accidents," in *International Handbook of Traumatic Stress Syndromes*, ed. J. P. Wilson and B. Raphael (New York: Plenum Press, 1993), 405–19; Liesel Ashley Ritchie, "Individual Stress, Collective Trauma, and Social Capital in the Wake of the Exxon Valdez Oil Spill," *Sociological Inquiry* 82, no. 2 (2012): 187–211.

19. R. Yehuda, J. Schmeidler, E. L. Giller, L. J. Siever, and K. Binder-Brynes, "Relationship Between Post-Traumatic Stress Disorder Characteristics of Holocaust Survivors and Their Adult Offspring," *American Journal of Psychiatry* 155, no. 6 (1998): 841–43; I. A. Kira, "Taxonomy of Trauma and Trauma Assessment," *Traumatology* 7, no. 2 (2001): 73–86; R. Lev-Wiesel, "Intergenerational Transmission of Trauma across Three Generations: A Preliminary Study," *Qualitative Social Work* 6, no. 1 (2007): 75–94.

20. Robert G. L. Waite, "The Holocaust and Historical Explanation," in *Genocide and the Modern Age: Etiology and Case Studies of Mass Death*, ed. Isidor Wallimann and Michael N. Dobkowski (New York: Greenwood Press, 1987), 165.

21. Marianne Hirsch, *Remembrance and Reconciliation: Encounters Between Young Jews and Germans* (New Haven, Conn.: Yale University Press, 2012).

22. Maria Yellow Horse Brave Heart, Josephine Chase, Jennifer Elkins, and Deborah B. Altschul, "Historical Trauma among Indigenous Peoples of the Americas: Concepts, Research, and Clinical Considerations," *Journal of Psychoactive Drugs* 43, no. 4 (2011): 282–90.

23. James Waller, *Becoming Evil: How Ordinary People Commit Genocide and Mass Killing*, 2nd ed. (New York: Oxford University Press, 2007), 174.

24. American Psychiatric Association, *Diagnostic and Statistical Manual of Mental Disorders: DSM-IV* (Washington, D.C.: American Psychiatric Association, 1994).

25. Harvey A. Barocas and Carol B. Barocas, "Manifestations of Concentration Camp Effects on the Second Generation," *American Journal of Psychiatry* 11, no.1 (1973): 6–14; Harvey A. Barocas and Carol B. Barocas, "Separation—Individuation Conflicts in Children of Holocaust Survivors," *Journal of Contemporary Psychotherapy* 11, no. 1 (1980): 6–14; Natan P. Kellermann, "Transmission of Holocaust Trauma—An Integrative View," *Psychiatry* 64, no. 3 (2001): 256–67; Natan P. Kellermann, "The Long-Term Psychological Effects and Treatment of Holocaust Trauma," *Journal of Loss and Trauma* 6, no. 3 (2001): 197–218.

26. Zahava Solomon, Moshe Kotler, and Mario Mikulincer, "Combat-Related Posttraumatic Stress Disorder among Second-Generation Holocaust Survivors: Preliminary Findings," *American Journal of Psychiatry* 145, no. 7 (1988): 865–68.

27. R. Krell, "Child Survivors of the Holocaust—Strategies of Adaptation," *Canadian Journal of Psychiatry* 38 (1993): 384–89; G. R. Leon, J. N. Butcher, M. Kleinman, A. Goldberg, and M. Almagor, "Survivors of the Holocaust and Their Children: Current Status and Adjustment," *Journal of Personality and Social Psychology* 41 (1981): 503–16; Peter Suedfeld, "Reverberations of the Holocaust Fifty Years Later: Psychology's Contributions to Understanding Persecution and Genocide," *Canadian Psychology* 41 (2000): 1–9.

28. Y. E. Danieli, *International Handbook of Multigenerational Legacies of Trauma* (New York: Plenum Press, 1998); W. G. Niederland, "The Survivor Syndrome: Further Observations and Dimensions," *Journal of the American Psychoanalytic Association* 29 (1981): 413–25.

29. Z. Solomon and E. Prager, "Elderly Israeli Holocaust Survivors during the Persian Gulf War—A Study of Psychological Distress," *American Journal of Psychiatry* 149 (1992): 1707–10.

30. https://twitter.com/neiltyson/status/695759776752496640.

31. Peter Mair, "Democracies," in *Comparative Politics,* ed. Daniele Caramani (New York: Oxford University Press, 2008).

32. Francis Fukuyama, *The End of History and the Last Man* (New York: The Free Press, 2006).

33. Pierre Hazan, "Transitional Justice after September 11: A New Rapport with Evil," in *Localizing Transitional Justice: Interventions and Priorities After Mass Violence,* ed. Rosalind Shaw and Lars Waldorf (Stanford, Calif.: Stanford University Press, 2010).

34. John Keane, *Violence and Democracy* (New York: Cambridge University Press, 2004).

35. See for example David C. Rapoport and Leonard Weinberg, "Elections and Violence," in *The Democratic Experience and Political Violence*, ed. David C. Rapoport and Leonard Weinberg (Oxford: Routledge, 2012).

36. John P. Crank and Linda S. Jacoby, *Crime, Violence, and Global Warming* (New York: Routledge, 2015).

37. See Rummel, *Death by Government.*

38. Jack Goldstone, Ted Robert Gurr, Barbara Harff, Marc A. Ley, Monty G. Marshall, Robert H. Bates, David L. Epstein, Colin H. Kahl, Pamela T. Surko, John C. Ulfelder, and Alan N. Unger, "State Failure Task Force Report: Phase III Findings," Science Applications International Corporation, September 30, 2000, accessed November 24, 2016, https://www.hks.harvard.edu/fs/pnorris/Acrobat/stm103%20articles/StateFailureReport.pdf.

39. Gary LaFree, and Andromachi Tseloni, "Democracy and Crime: A Multilevel Analysis of Homicide Trends in Forty-Four Countries, 1950–2000," *Annals of the American Academy of Political and Social Science* 605 (2006): 25–49.

40. Michael Mann, *The Dark Side of Democracy: Explaining Ethnic Cleansing* (New York: Cambridge University Press, 2005).

41. Gareth Evans and Mohamed Sahnoun, *Report of the International Commission on Intervention and State Sovereignty: The Responsibility to Protect* (Ottawa, Ontario: International Development Research Centre, 2001); Evans, *The Responsibility to Protect.*

42. Madeleine Albright and William S. Cohen, *Preventing Genocide: A Blueprint for U.S. Policymakers* (Washington, D.C.: United States Holocaust Memorial Museum Press, 2008).

43. See for example Sarah Sewall, Sally Chin, and Dwight Raymond, *Mass Atrocity Response Operations: A Military Planning Handbook* (CreateSpace Press, 2010).

44. For a good discussion see Gareth Evans, *The Responsibility to Protect.*

45. Richard J. Goldstone and Adam M. Smith, *International Judicial Institutions: The Architecture of International Justice at Home and Abroad* (New

York: Routledge, 2009); Kingsley Chiedu Moghalu, *Global Justice: The Politics of War Crimes Trials* (Stanford, Calif.: Stanford University Press, 2008); Geoffrey Robertson, *Crimes Against Humanity: The Struggle for Global Justice* (New York: Penguin Books, 2002); Howard Ball, *Prosecuting War Crimes and Genocide: The Twentieth-Century Experience* (Lawrence: University Press of Kansas, 1999); Gary Jonathan Bass, *Stay the Hand of Vengeance: The Politics of War Crimes Tribunals* (Princeton, N.J.: Princeton University Press, 2000).

46. Kathryn Sikkink, *The Justice Cascade: How Human Rights Prosecutions Are Changing World Politics* (New York: W.W. Norton & Company, 2011).

47. Sikkink, *The Justice Cascade*.

48. Priscilla B. Hayner, *Unspeakable Truths: Transitional Justice and the Challenge of Truth Commissions*, 2nd ed. (New York: Routledge, 2011); Eric Stover and Harvey M. Weinstein, eds., *My Neighbor, My Enemy: Justice and Community in the Aftermath of Mass Atrocity* (New York: Cambridge University Press, 2004); Mark Freeman, *Necessary Evils: Amnesties and the Search for Justice* (New York: Cambridge University Press, 2009); Pierre Hazan, *Judging War, Judging History: Behind Truth and Reconciliation* (Stanford, Calif.: Stanford University Press, 2010); Alexander Laban Hinton, ed., *Transitional Justice: Global Mechanisms and Local Realities After Genocide and Mass Violence* (New Brunswick, N.J.: Rutgers University Press, 2011).

49. Alexander Laban Hinton, "Introduction: Toward an Anthropology of Transitional Justice," in *Transitional Justice: Global Mechanisms and Local Realities After Genocide and Mass Violence*, ed. Alexander Laban Hinton (New Brunswick, N.J.: Rutgers University Press, 2011), 11.

50. Hans Köchler, introduction to *Global Justice or Global Revenge? International Criminal Justice at the Crossroads* (Vienna, Austria: Springer-Verlag Wien, 2003).

51. Köchler, introduction, 2.

52. Maureen S. Hiebert, "Do Criminal Trials Prevent Genocide? A Critical Analysis," in *Impediments to the Prevention and Intervention of Genocide*, vol. 9 of *Genocide: A Critical Bibliographic Review* (New Brunswick, N.J.: Transaction Publishers, 2013), 238.

53. Jan Willem Honig and Norbert Both, *Srebrenica: Record of a War Crime* (New York: Penguin Books, 1996); David Rohde, *Endgame. The Betrayal and Fall of Srebrenica: Europe's Worst Massacre since World War II* (New York: Farrar, Straus, and Giroux, 1997).

54. See for example Adam M. Smith, *After Genocide: Bringing the Devil to Justice* (Amherst, N.Y.: Prometheus Books, 2009).

55. Sewell Chan and Marlise Simons, "South Africa to Withdraw from International Criminal Court," *The New York Times*, October 21, 2016, accessed November 13, 2016, http://www.nytimes.com/2016/10/22/world/africa/

south-africa-international-criminal-court.html; Luckystar Miyandazi, Philomena Apiko, and Faten Aggad-Clerx, "Why an African Mass Withdrawal from the ICC Is Possible," *Newsweek*, November 2, 2016, accessed November 13, 2016, http://www.newsweek.com/icc-international-criminal-court-africa-gambia-south-africa-burundi-515870.

56. Jo Biddle, "Russia in New Blow to ICC as Court Urges Nations 'Don't Go,'" *Agent France Press*, November 16, 2016, accessed November 18, 2016, https://www.yahoo.com/news/russia-withdraws-signature-iccs-founding-statute-ministry-115751622.html.

57. Rasmus Heltberg, Paul Bennett Siegel, and Steen Lau Jorgenson, "Social Policies for Adaptation to Climate Change," in *Social Dimensions of Climate Change: Equity and Vulnerability in a Warming World,* ed. Robin Mearns and Andrew Norton (The World Bank, 2010), 259, accessed September 23, 2016, https://openknowledge.worldbank.org/handle/10986/2689.

58. Alex Alvarez and Ronet Bachman, *Violence: The Enduring Problem*, 3rd ed. (Thousand Oaks, Calif.: Sage, 2017).

59. Elijah Anderson, *Code of the Street: Decency, Violence, and the Moral Life of the Inner City* (New York: W. W. Norton and Co, 1999), 32.

60. William Alex Pridemore, "Poverty Matters: A Reassessment of the Inequality-Homicide Relationship in Cross-National Studies," *British Journal of Criminology* 51, no. 5 (2011): 739–72; Carol L. S. Trent and William Alex Pridemore, "A Review of the Cross-National Empirical Literature on Social Structure and Homicide," in *Handbook of European Homicide Research: Patterns, Explanations, and Country Studies*, ed. Marike Liem and William Alex Pridemore (New York: Springer, 2012); Travis C. Pratt and Francis T. Cullen, "Assessing Macro-Level Predictors and Theories of Crime: A Meta-Analysis," in *Crime and Justice: A Review of Research*, vol. 32, ed. Michael Tonry (Chicago: University of Chicago Press, 2015).

61. Elliott Currie, *The Roots of Danger: Violent Crime in Global Perspective* (New York: Oxford University Press, 2016).

62. Amy Chua, *World on Fire: How Exporting Free Market Democracy Breeds Ethnic Hatred and Global Instability* (New York: Anchor, 2004).

63. James Gilligan, *Preventing Violence* (New York: Thames and Hudson, 2001), 82.

64. United States Agency for International Development (USAID), "Vision for Ending Extreme Poverty," 2016, accessed January 6, 2017, https://www.usaid.gov/sites/default/files/documents/1870/Vision-XP_508c_1.21.16.pdf.

65. USAID, "Vision for Ending Extreme Poverty."

66. Jeffrey Sachs, *The End of Poverty: Economic Possibilities for Our Time* (New York: Penguin Books, 2006).

67. Paul Collier, *The Bottom Billion: Why the Poorest Countries Are Failing and What Can Be Done about It* (New York: Oxford University Press, 2007).

68. J. Andrew Plowman, *Climate Change & Conflict Prevention* (Washington, D.C.: National Intelligence University, 2014), 93.

69. Sebastian Junger, *Tribe: On Homecoming and Belonging* (New York: Twelve Books, 2016).

70. Junger, *Tribe*, 53–54.

71. Yuval Noah Harari, "Does Trump's Rise Mean Liberalism's End?" *The New Yorker*, October 7, 2016, accessed October 12, 2016, http://www.newyorker.com/business/currency/does-trumps-rise-mean-liberalisms-end.

72. Steven Pinker, *The Better Angels of Our Nature: Why Violence Has Declined* (New York: Viking, 2011).

73. Ian Morris, *War! What Is It Good For? Conflict and the Progress of Civilization from Primates to Robots* (New York: Farrar, Straus and Giroux, 2014).

74. Global Intelligence Council, *Global Trends: 2030: Alternative Worlds* (December 2012): 8, accessed November 13, 2016, https://www.dni.gov/index.php/about/organization/global-trends-2030.

75. W. B. Yeats, "The Second Coming," in *The Pocket Book of Modern Verse,* ed. Oscar Williams (New York: Washington Square Press, 1967).

INDEX

adaptation: capacity for, 27, 125; to climate, 139; failure of, 43; occurrence of, 29
The Age of Triage (Rubenstein), 132
Agier, Michel, 134
agriculture: climate uncertainty from, 38–39; dependency on, 126; DNA altered by, 49; drought influencing, 106; introduction of, 38; precipitation influencing, 25; support from, 38; water usage for, 98
air pressure, 12
alarmism, 2
Anasazi, 37; ancestry of, 43; evolution of, 38
Anderson, Elijah, 152
animal species: adaptation capacity of, 27; climate change impacting, 27–28
Antarctica, 23
Anthropocene, 18
anthropogenic change, 17, 154; consequences of, 21; natural world impacted by, 29
anti-Semitism, 54–56
aquifers, 99
Arctic freeze, 9
Armenian genocide, 54
Armstrong, Karen, 39
al-Assad, Bashar, 93
atmosphere, 11; carbon dioxide levels in, 19–20, 99, 140; dust in, 16; oceans'

interaction with, 15, 24
authoritarianism, 74, 77, 92, 124, 142, 148
Axis Rule in Occupied Europe (Lemkin), 52

Bandura, Albert, 83
Bangladesh, 7, 135–138
Barnosky, Anthony, 27
Bates, Robert, 76–77
Bauer, Wolfgang, 121
Beck, Aaron, 57
belief systems: borders and, 133; circumstances shaping, 84
Bering Land Bridge, 31
The Better Angels of Our Nature (Pinker), 157
Betts, Alexander, 126
biosphere, 12
The Black Book: Imbalance of Power and Wealth in Sudan, 110–111
bloodlands, 134
Border Patrol, 117
borders, 4; belief systems and, 133; changing of, 134; climate change, displacement and, 6–7, 115; climate change, violent conflict and, 132, 134–135; European, as dangerous, 121; frontier mentality of, 132; hardening of, 154; Mexico and U.S., 115–117; militarization of, 118, 130; oceans as, 121; reinforcement of, 131;